T0366686

ROSALEEN DUFFY

SECURITY AND CONSERVATION

The Politics of the Illegal Wildlife Trade

Yale
UNIVERSITY
PRESS
NEW HAVEN AND LONDON

Research for this book was supported by European Research Council (ERC) Advanced Investigator Grant # 694995 entitled *BIOSEC: Biodiversity and Security, Understanding Environmental Crime, Illegal Wildlife Trade and Threat Finance.*

Published with assistance from the Louis Stern Memorial Fund.

Yale University Press books may be purchased in quantity for educational, business, or promotional use. For information, please e-mail sales.press@yale.edu (U.S. office) or sales@yaleup.co.uk (U.K. office).

Set in New Aster and Syntax LT type by IDS Infotech, Ltd.
Printed in the United States of America.

Library of Congress Control Number: 2021944802
ISBN 978-0-300-23018-5 (hardcover : alk. paper)

A catalogue record for this book is available from the British Library.

This paper meets the requirements of ANSI/NISO Z39.48-1992
(Permanence of Paper).

10 9 8 7 6 5 4 3 2 1

For Feargal and Oisín

CONTENTS

PREFACE

"Look at this, it has now become an industry," remarked a former colleague, with a wry smile. We were standing in the center of Battersea Evolution, the venue for the London Conference on the Illegal Wildlife Trade in October 2018. The two-day event was quite a spectacle with more than one thousand participants. There was a secure mezzanine floor for heads of government and official state delegations, and there were black-suited bodyguards everywhere. Booths advertising the latest campaigns and innovations to tackle the illegal wildlife trade ringed the ground floor. There were a series of rhino statues, painted with different striking designs to draw attention to their plight. A bespoke sculpture of a person stood at the center of an elephant's head, holding on to the tusks.[1] The heavy presence of the British Army, tech companies, and private security services was noticeable among the usual attendees, such as WWF, TRAFFIC International, and National Geographic. The conference had the feel of a jamboree mixed with a trade show. The plenaries and side events were the entertainment, but the real deals were being made in the mezzanine floor, around the display booths, and at the coffee stations. The delegations from governments, NGOs, and the private sector were keen to strike deals for new collaborations and, crucially, for new streams of funding.

It was all very different from the first London conference on the illegal wildlife trade in 2014, where there was a closed heads of government meeting, with a linked two-day event in a much smaller room at the Zoological Society of London. That venue held around 250 people; everywhere I turned I could see the important (white) figures of the conservation scene: Jane Goodall, Will Travers, Iain Douglas-Hamilton. They sat beside me, walked to the coffee breaks with me, asked questions, and engaged in informal discussion. Prince William showed up at both conferences—he was a surprise visitor on day two—in 2014; in 2018 he was a scheduled keynote speaker on day one.[2] After his address in 2018, Stephen Corry of Survival International accused him of promoting a white saviour complex, expressing colonial attitudes toward conservation and showing a film in which white conservationists were portrayed as the caring experts while Africans were cast as grateful recipients.[3] In 2014, it was the early days of thinking about wildlife conservation as a security issue. At that conference, the military, intelligence services, and tech companies were barely noticeable, if they were there at all. What had changed, and why?

This book is the story of that shift, why it happened, and what it means for wildlife conservation. This is not to claim the security turn is the only response by conservationists to the illegal wildlife trade. There are also important attempts to engage in demand reduction and community-based approaches to tackle poaching and trafficking, and these also need to be seen in the wider context of conservation practices which cover a broad range of approaches, including participatory and community-oriented policies, working in partnership with Indigenous peoples, and more neoliberal schemes, such as payments for ecosystem services and biodiversity offsetting.[4] Further, geographical location, risks to wildlife, which donors and NGOs are engaged, and the historical trajectory of conservation in the area, among other concerns, factor into approaches to conservation.[5] The shift toward security in conservation does not mean that other options have been completely abandoned; rather, security-oriented approaches coexist alongside others, but they have become more prominent in policies to tackle illegal wildlife trade especially. Therefore, in this book I specifically focus on responses that move toward thinking of conservation as a security issue. The "security turn" in wildlife conservation, more broadly, has indeed become an industry in its own right. Millions of dollars have poured in from such donors as the Global Environmental Facility and USAID and from national governments, private foundations, philanthropists, corporations, and the general public (see chapter 1). Mainstream conservation NGOs are the key beneficiaries of these new sources of funding, but so are a raft of new entrants to the conservation scene: private military companies, intelligence services, risk analysts, surveillance technology developers, and drone manufacturers. It is driving a new phase in conservation in which the conservation and security sectors are becoming more integrated and are shaping each other in complex ways. This poses the question: how did we get here?

In this book I aim to answer that question by examining the growing convergence of security and conservation. This is not another book about the growth of the illegal wildlife trade or its impact on particular species—there are substantial treatments of these topics already.[6] My focus here is: why did the conservation sector turn toward security as the key response for tackling the illegal wildlife trade, how was it financially and politically supported, and what effects has this had on conservation more widely. The illegal wildlife trade has gained international prominence since the sudden rises in poaching of elephants and rhinos from approximately 2008 to 2010. This increase forged a renewed sense of urgency in conservation, a call to *do something* to tackle poaching and trafficking before species are driven to extinction. This sense of urgency has increased as a result of the COVID-19 pandemic. It is probable that the source of the pandemic was a wildlife market, which has heightened concerns about the impacts of the illegal trade on human health and on biodiversity. There have been high-profile calls from conservation and animal welfare organizations for immediate bans on legal wildlife markets as a means of preventing future pandemics. This urgency has produced a series of important conceptual and practical shifts in conservation, which demand a more thorough investigation and analysis.

To consider these issues, I take a political ecology approach, which focuses on socioecological relations. Political ecologists have been at the forefront of critiques of the social, political, and economic impacts of conservation, especially the experiences of more marginalized and vulnerable communities. However, to date, political ecologists have not engaged sufficiently with understanding how human-environment relations underpin and flow through certain ways of thinking about global security (for more detail, see chapter 1). Using political ecology, it is possible to analyze the ways in which security and conservation are becoming more integrated. It is rapidly, and fundamentally, restructuring our relationships with nature in ways that will continue to shape interactions between people and wildlife in the longer term. By aligning more closely with security to tackle the illegal wildlife trade, conservation has shifted in a new direction, reshaped by the approaches, techniques, and practices from the security sector. This shift could cut out options for conservation in the future.

The aims of conservation and of security are very different. The central or urgent issue that needs to be addressed in the illegal wildlife trade is the prevention of species declines and losses as a result of unsustainable extraction and trade. By contrast, the objective of security, in traditional terms, is to prevent or tackle threats to global and national stability. The significant shift in responses to the illegal wildlife trade is that the conservation aims are subsumed under those of security, such that nature becomes reconfigured so as to contribute to global security.

This critical shift around the illegal wildlife trade has been facilitated by the argument that the trade constitutes a security threat. This thinking is underpinned by particular ways of framing illegal wildlife trade as a security issue. Although there are some historical antecedents (see especially chapters 2, 3, and 7), the convergence of security and conservation thinking represents a new phase. This is underpinned by framing the illegal wildlife trade as a serious crime and as a source of finance for organized crime, armed groups, and terrorist networks that threaten national and global stability. Conservation NGOs have been quick to take up this theme and use it to draw attention to their work and as a means of gaining new streams of funding for their activities (see chapters 2 and 4).

These dynamics have produced particular sets of practices in conservation. Approaching the illegal wildlife trade as a security issue prompts a security-type response. This shift toward security thinking has encouraged the entry into conservation of security sector actors, who are often seeking new markets for their training packages, weapons, and surveillance systems. It deepens the engagement of private sector operators, especially those offering technological solutions for tackling illegal wildlife trade. Furthermore, in places where there is intense concern that poaching will drive the extinction of iconic species, it has facilitated the militarization of conservation (see chapter 3).

I set this out in the following chapters, each of which offers an in-depth exploration of a specific theme in the security turn in conservation. Throughout this book I draw on interviews with people working for conservation NGOs, donor agencies, intelligence and technology businesses, private military companies, intelligence agencies, and government departments. The individuals are anonymized in accordance with ethics guidance of the European Research Council and University of Sheffield; guaranteeing interviewees anonymity and confidentiality allowed them to talk candidly about their work and about the integration of security and conservation. This provided important firsthand insights into the rationale for developing security-oriented approaches, as well as the challenges and divisions within the conservation community over militarization, reliance on new technological solutions, and use of trained intelligence operatives.

It is reasonable to ask, why should it concern us if the illegal wildlife trade does indeed threaten species with extinction and contributes to national and global insecurity? In this book, I show why it is a matter of concern. The security-oriented approach focuses on the roles of organized crime networks, armed groups, terrorist networks, and corruption. It gives rise to such solutions as intelligence gathering, enhanced law enforcement, or reliance on technologies. All of these are important and each can play a role in tackling the illegal wildlife trade. However, these approaches do not address the social, political, and economic complexities that fundamentally drive and sustain the illegal wildlife trade; rather they obscure this important wider context and the multiple steps that lead to the moment when rangers encounter poachers, ivory is intercepted at an airport, or stolen cactus ends up in a collector's home. In this book I focus specifically on how the sense of urgency and concern surrounding species losses and the illegal wildlife trade has prompted the conservation community to turn toward security-oriented approaches. My aim is to explore the (often unintentional) consequences of the positive motives to save species—essentially what strategies are chosen and why, what benefits and harms do these strategies distribute, and with what consequences for people and for wildlife.

Rhino display at the London Conference, 2018. Photo credit: Lucy Dunning.

Opening panel discussion at the 2018 London Conference, with the #EndWildlife-Crime slogan in the background. Photo credit: Lucy Dunning.

ACKNOWLEDGMENTS

This book represents a very long journey. I began thinking about it in 2011 when the arguments for seeing illegal wildlife trade as a global security threat first started to emerge. I was being invited to talk about my work to a new range of actors who had not really seemed interested in it previously: security think tanks, members of the military, and technology companies. I noticed that the usual academic and practitioner events about wildlife conservation that I attended increasingly included security companies demonstrating drones or intelligence analysts talking about how to build information networks. Conservation meetings were filled with a new language of situation rooms, intel gathering, weapons training, and working with assets. On one level it fascinated me, but on another it meant that I was propelled into a world in which I felt profoundly uncomfortable and where I did not feel I belonged. At the intellectual level it challenged me to think about the research I was undertaking and how research and evidence at the intersection of conservation and security can be operationalized in ways that present ethical and moral dilemmas. It is exciting to be working on research that is frequently in the media spotlight, but at the same time it can be daunting, and a little exposing, to publicly contest dominant narratives that present simplistic relationships between conservation and security. It becomes especially difficult when these claims are repeated in the mainstream media by powerful interests and organizations. In this sense, this book's journey to publication has been at times a fraught one as the arguments at its core have not always been ones that people have wanted to hear—or have others hear. Negative responses, though, served to increase my determination that these arguments need to be made. For these reasons, this book has been the most challenging one of my career to research and to write. Then, in the final stages, the world was overtaken by the emergence of COVID-19.

I had just returned from a brief trip to ATREE in Bangalore, where Dr Nitin Rai was a most generous host, and was about to travel on to Rwanda for a workshop on environmental peacebuilding, organized by Dr Elaine (Lan Yin) Hsiao at the University of Rwanda. I was so looking forward to the event and meeting Elaine and her colleagues and students in Kigali. But suddenly news of the spreading virus went from something distant to an immediate crisis that changed everything. Flights were cancelled, borders were closed, and new restrictions on meetings were announced. The United Kingdom seemed to be sleepwalking into disaster while the rest of the world locked down; it was not until 21 March 2020 that the UK government finally decided to go into lockdown too. Our son's school was closed; I wasn't allowed into my office at the university; we could not visit family; and like everyone else, we went outside only for exercise and essential shopping. Our dog Milo loved it—as he got loads of lovely walks and having everyone at home was bliss. It was difficult to continue with any kind of normal work when our world suddenly shrank down to the boundaries of our house. But we were among the most fortunate ones—and we counted our blessings.

This book was completed in this context. It took several years to research and write, and it would not have been possible without my wide network of support. I extend my sincerest thanks to my editor Jean Thomson Black, who was always patient and supportive. Jean's advice and wise counsel was crucial, and she has remained a champion of this project. The research for this book was funded by a very generous Advanced Investigator Award from the European Research Council (grant number 694995). This grant funded the four-year BIOSEC project on Biodiversity and Security, Understanding Environmental Crime, Illegal Wildlife Trade and Threat Finance (https://biosec.group.shef.ac.uk/). It involved a fantastic team, and the project itself has been a career highlight for me. I am especially grateful to team members Ruth Wilson, Lucy Dunning, Dr. Hannah Dickinson, Dr. Laure Joanny, Dr. Anh Vu, Dr. Teresa Lappe-Osthege, Dr. Sarah Bezan, Dr. Jared Margulies, Dr. George Iordachescu, and Dr. Francis Massé. They fostered a team dynamic of friendship, camaraderie, and support. They read and commented on various parts of the book and were instrumental in helping me to develop aspects of the argument; they also provided some of the photographic images included. Members of the BIOSEC project advisory board generously gave up their time to advise us and shape the direction of the research; thanks to Dr. Dilys Roe, Sabri Zain, Professor EJ Milner-Gulland, Professor Maano Ramutsindela and Professor Tor Arve Benjaminsen; and thanks to Professsor Bram Büscher and Professor Libby Lunstrum for reading and commenting on the book proposal and chapters. The wider fellowship of the BIOSEC project and the Political Ecology Reading Group have also provided excellent constructive comments on various sections of this book, particular thanks go to Dr. Adeniyi Asiyanbi, Dr. Esther Marijnen, Dr, Judith Krauss, Dr. Elaine (Lan Yin) Hsiao, Dr. Brock Bersaglio, and Dr. Charis Enns. They have been central to my finishing this book, and I cannot thank them enough. Professor Dan Brockington has been a constant source of support for fifteen years now, since we first worked together at Manchester University. He is always there, ready with calming words, excellent advice, and a joke to cheer me up when things get tough. He is quite simply the best colleague and friend anyone could ask for.

The Department of Politics and International Relations at the University of Sheffield was an excellent place to write this book. I am grateful to three Heads of Department who supported me and the BIOSEC project: Professor Nicola Phillips, Professor Andy Hindmoor, and Professor Ruth Blakeley. Each listened carefully and provided clear and constructive advice about the project, the bumpy road to publication, and, most recently, what I should do next. I am also grateful for the intellectual input and friendship provided by colleagues who read papers and chapters and gave advice; some of them may not be aware that a casual conversation in the corridor sparked a train of thought that ultimately shaped the book. Thank you Sarah Beddow, Jennifer Watson, Dr. Jonna Nyman, Dr. Jo Tidy, Dr. Alasdair Cochrane, Dr. Ross Bellaby, Dr. Helen Turton, Dr. Judith Verweijen, Dr. Simon Rushton, Professor Charlie Burns, Professor Jan Selby, Dr. Tom Johnson, Dr. Josh Milburn, Dr. Anastasia Shesterinina, Dr. Burak Tansel, Professor Allister MacGregor, Dr. Perla Polanco Leal, Dr. Peter Sands, and Dr. Lisa Stampnitzky.

Throughout the research process I have benefited so much from academics, conservationists, and security specialists who generously answered questions or helped to facilitate meetings. I thank Professor Tanya Wyatt, Professor Shannon O'Lear, Dr. Meredith Gore, Dr. David Roberts, Keith Somerville, Charles Jones Nsonkali at Okani, Dr. Rosie Cooney at IUCN-SuLi, Dr. Cathy Dean and Dr. Susie Offord-Wooley at Save the Rhino, Paul de Ornellas at ZSL/WWF UK, Rob Parry-Jones at WWF-International, Rebecca Drury and Joanna Elliot at FFI, Sophie Grig and Freddy Weyman at Survival International, Maud Salber and Simon Counsell at Rainforest Foundation UK, Douglas Carpenter at EEAS, Dr. Matt Luizza and Dr. Daphne Carlson-Bremer at USFWS, John Waugh at LLC, and Hasita Bhammar at the World Bank. Several people cannot be named for reasons of anonymity, but you know who you are and I am deeply grateful for your candour. Any errors or misinterpretations are, of course, mine.

I am also grateful for the love and support of family. It is fair to say that we have had a few very tough years. While I was researching and writing this book, we lost our Mum, Dad, and then our brother Terry, after very long battles with debilitating illnesses. Sadly, grieving became everyday life. Thank you Carmel, Mary, Simon, Dennis, Katherine, Liza, Annabel, Alex, Peter, Joanna, Thomas, Orlaith, and Zeke. COVID-19 prevented us from meeting to grieve together, but when we can be together we will raise a glass, a smile, a tear. My husband's family—Roisín, Gerry, Niall, and Geraldine—have been there with kind words and support during these difficult times.

Finally, this book could not, and would not, have been written without my husband, Professor Feargal Cochrane, who was writing his own books at the time. He helped me keep a sense of perspective, was always a wise sounding board, and picked me up when life was at its worst. For several years he undertook a grueling commute, separated from us for several days each week, so that we could move to the Peak District and I could join Sheffield University. That changed in 2020 when he chose family life and he was home. Little did we realize just how different 2020 would be. Suddenly confronted with homeschooling our son Oisín, we have learned the new techniques in maths, and despite the fact that as academics our job is to write, we only just found out what a fronted adverbial is (and no, our lives are not richer for that). Oisín is a fantastic kid, bursting with life and energy—he is the center of our lives. Having academics as parents has meant that he also likes to share knowledge and would proudly go in to his primary school to explain Brexit, the Irish Backstop, or why we should care about the illegal wildlife trade. Lockdown has been hard for him too: running about with friends and cousins has been replaced with gaming and chatting online. But he remains positive and is a lesson to us all.

As I complete this book, I hope that we are on the road out of the pandemic, which shone a light on the problematic relationships between humans and the natural world. Some refer to our current era as "the Anthropocene"—a world shaped by "human" impacts. But this obscures the structural conditions that shape the world and make us all equally responsible, when actually we are not. It is too easy to place blanket blame on all humanity for the pandemic, and for other crises like climate change, biodiversity losses, inequality, pollution, and deforestation. Instead, these crises are structured by the current global system that produces spectacular wealth for a minority but also keeps so many in vulnerable, precarious, and marginalized lives. It is these inequalities that sustain the illegal wildlife trade.

My ambition in this book to show that we need to address and tackle these underlying drivers to produce a fairer and sustainable future and to prevent species losses as a result of illegal and unsustainable trade.

Rosaleen Duffy
Sheffield University

ABBREVIATIONS

ADMADE	Administrative Management Design Programme
AEFF	African Environmental Film Foundation
ASEAN	Association of Southeast Asian Nations
CAMPFIRE	Communal Areas Management Plan for Indigenous Resources
CAR	Central African Republic
CBMC	Community-Based Militarized Conservation
CBNRM	Community-Based Natural Resource Management
CI	Conservation International
CITES	Convention on the International Trade in Endangered Species
CoP	Conference of Parties
DRC	Democratic Republic of Congo
DEFRA	Department of Environment, Food and Rural Affairs (UK)
DfID	Department for International Development
DG-ENV	Directorate General Environment of the European Commission
EAGLE	Eco Activists for Governance and Law Enforcement Network
EAL	Elephant Action League

ECOSOC	UN Economic and Social Council
EIA	Environmental Investigation Agency
EFFACE	European Union Action to Fight Environmental Crime
EMPACT	European Multidisciplinary Platform Against Criminal Threats
END	Eliminate Neutralize Disrupt (END) Wildlife Trafficking Act
ETIS	Elephant Trade Information System
ETF	Ecological Task Force
EUROJUST	European Union Agency for Criminal Justice Cooperation
EUROPOL	European Union Agency for Law Enforcement Cooperation
EU-TWIX	EU Trade in Wildlife Information eXchange
FARC	Fuerzas Armadas Revolucionarias de Colombia
FARDC	Forces Armées de la République Démocratique du Congo
FATF	Financial Action Task Force
FCO	Foreign and Commonwealth Office
FFI	Flora and Fauna International
FIU	Financial Intelligence Unit
GEF	Global Environment Facility
GRAA	Game Rangers Association of Africa
GTI	Global Transparency Initiative
IAPF	International Anti Poaching Foundation
ICC	International Conservation Caucus
ICCF	International Conservation Caucus Foundation
ICCN	Institut Congolais pour la Conservation de la Nature
ICCWC	International Consortium on Combatting Wildlife Crime
ICRC	International Committee of the Red Cross
IFAW	International Fund for Animal Welfare
INTERPOL	International Criminal Police Organization
IUCN	International Union for the Conservation of Nature

IUU	Illegal, unregulated, and underreported
IWT	Illegal Wildlife Trade
KCAA	Kenya Civil Aviation Authority
KPRs	Kenya Police Reserves
KWS	Kenya Wildlife Services
LATF	Lusaka Agreement Task Force
LEMIS	Law Enforcement Management Information System
LRA	Lord's Resistance Army
M2M	Military to Military
MEP	Member of the European Parliament
MIKE	Monitoring Illegal Killing of Elephants
MOSSAD	HaMossad leModi'in uleTafkidim Meyuḥadim
MPA	Marine Protected Area
NRT	Northern Rangelands Trust
OECD	Organization for Economic Cooperation and Development
PDSA	People's Dispensary for Sick Animals
PMCs	Private Military Companies
PWCF	Prince of Wales Charitable Foundation
REDD+	Reducing Emissions from Deforestation and forest Degradation
RFUK	Rainforest Foundation UK
RUSI	Royal United Services Institute
SANParks	South African National Parks
SAS	Special Air Services
SDSR15	2015 Strategic Defence and Security Review
SMART	Spatial Monitoring and Reporting Tool
SOCTA	Serious and Organized Crime Threat Assessment
SPLA	Sudan People's Liberation Army
SPLM	Sudan People's Liberation Movement
TNC	The Nature Conservancy
TRAFFIC	Trade Records Analysis of Flora and Fauna in Commerce
UAVs	Unmanned Aerial Vehicles
UfW	United for Wildlife

UNCTAD	United Nations Conference on Trade and Development
UNEP	United Nations Environment Program
UNODC	United Nations Office on Drugs and Crime
UNPKOs	United Nations Peace Keeping Operations
UNTOC	United Nations Convention against Transnational Organized Crime
USAID	United States Agency for International Development
USFWS	US Fish and Wildlife Service
USGIF	US Geospatial Intelligence Foundation
WD4C	Working Dogs for Conservation
WENs	Wildlife Enforcement Networks
WCMC	World Conservation Monitoring Centre
WCN	Wildlife Conservation Network
WCO	World Customs Organization
WCS	Wildlife Conservation Society
WJC	Wildlife Justice Commission
WWF	World Wide Fund for Nature
WWF-US	World Wildlife Fund-US
ZSL	Zoological Society of London

1 CONSERVATION AND SECURITY CONVERGE

Political Ecologies of the Illegal Wildlife Trade

Conservation is changing. We are accustomed to the notion that it is directed by scientists wielding notebooks, running computer models, watching wildlife behaviors, or collecting plant samples. The rapid growth in poaching and trafficking of some of the world's most iconic species, however, has generated a sense of urgency, and conservation, and conservationists, now look very different. In some places the notebooks, computers, and sample jars have been replaced with weapons, hi-tech surveillance systems, and intelligence networks. It has become harder and harder to distinguish between conservationists and the people and practices normally found in the security sector. Scientists are being replaced by security operatives trained in intelligence gathering, use of weapons, and surveillance techniques. The object of their gaze is not the plants and animals, but the humans that might threaten them. This change signals a key shift. The attention of conservationists is moving away from a focus on ecological monitoring and data gathering, toward a fuller focus on people (as individuals or networks) who are defined as a threat to wildlife. This is most prominent in responses to the illegal wildlife trade, but it is also reshaping conservation more broadly.

The illegal wildlife trade has gained international attention since poaching of elephants and rhinos began to increase in the late 2000s. By 2019 there was some good news about slowing rates of poaching, but then the eruption of COVID-19 brought new concerns and a focus on the risks posed by the illegal wildlife trade. Conservation and animal welfare NGOs, among others, seized on the pandemic as evidence for the need for tougher regulations of all wildlife markets, whether legal and regulated or illegal and unregulated, in the interests of public health. This has forged the current sense of urgency that lies behind the rise of security in conservation. Previously we had been taking these steps to prevent extinctions; now we must act because of significant threats to human health and well-being.

The new sense of urgency has produced a series of important conceptual and practical shifts in conservation. It has opened an important space for security thinking, practices, and practitioners to move into conservation. These changes have encouraged and facilitated the entry of militaries, private sector security companies as well as intelligence and risk companies into conservation, and they demand further investigation. Further, the urgency has allowed these groups to generate new rationales for their activities and to justify greater allocation of resources to security as a means of saving species. Scrutiny of how funds are used, where they are used, and with what consequences is limited by the very same security precautions that necessitate the move toward partnering with militaries and security companies in first place.

There is an element of opportunism at work for both conservation and the security sectors. Each is entrepreneurial. For the security sector, conservation provides new markets and testing grounds. For conservation, security brings important new streams of funding, a great deal of public attention, and policy commitments from some of the world's most powerful actors. The two forces are combining to create more security-oriented responses to the illegal wildlife trade, which are also reshaping approaches in conservation as a whole. The shift toward security does not mean that other ways of doing conservation have been completely abandoned. Security approaches coexist alongside others that are anchored in community-based and neoliberal strategies, but the integration of security with conservation is more prominent in policies to tackle illegal wildlife trade.

This shift raises some key questions. Where do the *aims* of conservation and security diverge, and why it is so interesting and concerning that they are being blended together? In what ways have the specific *framings* of the illegal wildlife trade as a serious organized crime, or source of finance for armed groups, facilitated and deepened the integration of security and conservation? And how do these lead to a new set of *practices* on the ground, changing conservation such that policy has become folded into security strategies? What is the emerging debate on political ecologies of conflict and conservation, which can act as an initial foundation for developing a political ecology approach to understanding the integration of conservation and security? And, what is the role of funding in developing and supporting a more security-oriented approach to conservation?

CONVERGENCE

We can see the convergence of conservation and security in the expanding number and range of people and private businesses that offer training for and implementation of anti-poaching in sub-Saharan Africa (see chapter 7). They offer their skills, modes of thinking, and approaches from experiences in conflicts across the globe, especially from U.S.-led international interventions in the Middle East. In April 2017, the Game Rangers Association of Africa (GRAA) issued a statement on the use of security, military personnel, and tactics in training Africa's rangers. The GRAA noted its concern about the shifting scene in conservation: "Military personnel, military veterans, and security contractors from beyond Africa's borders are becoming increasingly involved in ranger training across our continent. Intentions in some instances may be noble but there are mounting concerns that need to be noted by the ranger community in Africa."[1]

Among other concerns GRAA noted the lack of coordination, lack of understanding of the operating environment, and need for ecological sensitivity among groups offering training and enforcement. The GRAA pointed to the lack of knowledge of the legal frameworks in which rangers operate, a lack of proper vetting, and profiteering by military equipment manufacturers. These concerns were echoed by a senior U.S. government official with extensive experience of working on wildlife and security issues in Africa. In discussions with me the official drew attention to the ways that the engagement of private military companies in conservation in Africa were changing: "Ecomercenaries? It was growing in Africa, it's starting to find a better level. There have been sufficient examples of the unintended consequences and downsides to dampen down the initial enthusiasm for using experienced paramilitaries. . . . Now it is much more refined, it is more training oriented. Instead of being a large number of foreigners coming in wielding rifles, it's more subtle, training and accompanying."[2]

These security-oriented organizations do not just provide practical training; they bring with them broader attitudes and approaches as part of specialist knowledge and support. It is a growing practice, and it is more than a short-term response to an emergency situation: it is changing the nature of conservation in those areas in the longer term.

It is certainly the case that there are many organizations, spanning a range of practices and military backgrounds, that have moved into the conservation arena. They include (but this is in no way an exhaustive list) Veterans Empowered to Protect African Wildlife (VETPAW), Veterans4Wildlife, Chengeta Wildlife, International Anti Poaching Foundation (IAPF), Ecological Defense Group, Inc (EDGE), and Maisha Consulting. Each one aims to use a range of military and intelligence skills to assist in tackling the illegal wildlife trade. Beyond this, conservation NGOs, including Wildlife Conservation Society, African Parks Foundation, WWF-International, Zoological Society of London, among many others have begun contracting individual advisors with a military, intelligence, or policing background to advise them on how to tackle the illegal wildlife trade. This is not a specific criticism of these organizations. My interest here is in exploring further why and how this interaction and integration has developed, and how is it shaping responses to the illegal wildlife trade and reshaping some aspects of broader conservation practice.

How did this situation develop when the aims of security and conservation seem to differ? In chapter 2 I explain further the rationale for and means through which the illegal wildlife trade has become discursively defined as a security issue. Here, however, it is useful to set out briefly the main arguments for approaching the illegal wildlife trade as a security issue.

First, there is the argument that the illegal wildlife trade is a criminal activity, often involving organized crime networks. It is therefore appropriate, or even essential, to respond with greater levels of law enforcement to uphold the rule of law.[3] Such responses include the use of intelligence gathering, informant networks, and more active forms of policing.

Second, conservation activities and programs that operate in areas of armed conflict face specific challenges. In places marked by armed conflict, especially where the state is effectively absent, conservation agencies (public or private) operate in extremely difficult circumstances because their staff, as well as the wildlife, are at risk of attacks by armed groups and can struggle to maintain daily operations. In those circumstances, conservationists can feel they have no other option but to defend themselves and their programs via security-oriented practices including more militarized methods.[4]

Third, there is the argument that poachers are becoming more heavily armed and organized, and therefore the only way to respond effectively is through the use of more forceful methods, including arming rangers.[5] Community-based approaches, for example, are perceived not to work if those involved in hunting and trafficking are armed, organized, and interested in generating large profits for criminal networks.

Fourth, concern for future generations heightens the sense of emergency. Proponents of militarized conservation often present the use of force as a noble or heroic quest to save species.[6] In effect, conservationists have no option but to act now in order to save wildlife for future generations. This thinking is underpinned by the sentiment that more security-style responses and especially more forceful ones are an option of last resort.

These are strong arguments in favor of turning toward a security-oriented response. Where there is an emergency situation, conservationists need to act before it is too late. Armed, militarized responses can be faster, and they appear to be more effective than community-based approaches. In such challenging circumstances criticism of these last-resort options is often very unwelcome; it can be regarded as naïve or as an unhelpful distraction from the urgent challenges faced by conservation staff on the ground, especially rangers.[7]

Nevertheless, it is important to examine these shifts in the practices of conservation and subject them to careful scrutiny. It is an essential part of developing effective and socially just policies for the future and analyzing who benefits, who is disadvantaged, and why. My focus in this book is to examine the ways conservation is becoming integrated with more traditional understandings of security which focus on threats to national and global security. Conservation is at an important crossroads, a tipping point even, and so this book is a timely critical reflection at a significant juncture.

ILLEGAL WILDLIFE TRADE AND HUMAN SECURITY

There are important human security dimensions to the illegal wildlife trade. Biodiversity underpins human health and well-being, and any threat to biodiversity presents a threat to our very ability to survive. For example, the 2019 *Global Assessment Report* by Intergovernmental Science-Policy Platform on Biodiversity and Ecosystem Services (IPBES) indicated that more than two billion people rely on wood fuel to meet their primary energy needs, an estimated four billion people rely primarily on natural medicines for their health care, and more than 75 percent of global food crop types (fruits, vegetables, and cash crops, such as coffee, cocoa, and almonds) rely on animal pollination. Yet an estimated one million species are threatened with extinction.[8] Conservationists (among others) have repeatedly sounded the alarm that these losses will have a negative effect on human well-being and may even pose an existential threat to human life on earth. Debates on illegal wildlife trade and its contribution to biodiversity declines and losses need to be seen against this wider context.

Efforts to understand the links between environmental change—including illegal wildlife trade—and human security is a rapidly developing field, one that first emerged in the late 1990s; it has shifted the focus toward thinking about the security of individuals or populations to respond to threats to their basic needs and rights, including international protection of people from harms enacted by states on their own populace.[9] The concept of human security is very broad and therefore difficult to define in a precise way. O'Brien and Barnett, drawing on Amartya Sen's capabilities approach, define human security as a condition in which people and communities have the capacity to respond to threats to their basic needs and rights and thus can live with dignity.[10] There has been a growing realization that the illegal wildlife trade can shape the ability of people to meet their everyday food needs. The dynamic interactions between illegal wildlife trade and human security are fourfold.

First, illegal wildlife trade can negatively affect the ability of some subsistence and forest-dependent communities to meet their basic needs. In areas where wildlife is sourced, poaching and trafficking can deprive some communities of important sources of food, which may be one of their few sources of protein.[11] For example, forest-dependent peoples, such as the Baka, Aka, Bagyeli, Bakola, and Batwa in the Congo Basin, have traditionally engaged in hunting and fishing to meet their protein needs; poaching and trafficking wild caught meat for urban or external markets removes that resource. Further, the establishment of national parks and wildlife laws, often originally designed and implemented under colonial rule (but maintained after independence), criminalized that hunting. For many, consumption of wildlife is critically important for day-to-day survival, and increasing levels of enforcement of those laws have led to malnutrition in some communities.[12] Indeed, Jerome Lewis eloquently explains how the development of conservation initiatives in Congo were one of the key factors in impoverishing BaYaka communities that, for generations, had depended on forest resources to live; once cut off from the forests by conservation initiatives, development schemes, and logging they have become increasingly impoverished and dependent on precarious and low paid wage labor.[13]

Second, the removal of valuable wildlife by illegal wildlife trade can deprive communities, the private sector, and governments of important sources of income (in cases where wildlife is traded as food, is the source of tourism revenues, or provides income from trophy hunting). CITES states, for example, "Besides generating significant losses in assets and revenues for many developing countries, the theft of and illegal trade in natural resources potentially threatens the livelihood of rural communities, impacts upon food security, and risks damaging whole ecosystems."[14]

Third, it is commonly argued that poaching is related to poverty.[15] Conservationists often argue for more effective involvement of the rural poor in both development and conservation projects as a means of tackling poaching.[16] Arguments that illegal wildlife trade impacts negatively upon human security tend to rely on a very narrow, predominantly economic, definition of poverty.[17] Poverty, though, is more than economic deprivation; it encompasses concerns about status, prestige, and the ability to shape one's own future and lead a dignified life. Although Roe et al. point out that illegal wildlife trade is central to the livelihood strategies of some of the poorest communities in the world,[18] for others illegal wildlife trade is more than just a subsistence strategy. South Africa's rhino-poaching crisis, for example, is often attributed to poverty in Mozambique, identified as a significant "problem state" at CITES CoP16 in 2013.[19] However, following extensive research in the areas of Mozambique that border Kruger National Park, Lunstrum and Givá argue that it is not poverty per se that drives poaching but economic inequality. The lack of other job opportunities (in Mozambique and as migrant labor in South Africa) and an inability to access government financial support creates a context in which rhino poaching is attractive (despite the risks) because it can bring enormous and instant wealth. But people engage in the poaching economy for a range of economic reasons: to improve the lives of their families, to amass personal wealth, and to engage in conspicuous and unsustainable consumption.[20] Therefore, it is clear that poaching is more than just a matter of (narrowly defined) economic poverty. People who engage in poaching do not conform to the stereotype of the greedy criminal who cares little for the animals they kill. Rather, the drivers of poaching are multi-layered and complex; they relate to lack of opportunity, money, status, and wealth, as well as conspicuous consumption and a desire to gain respect.

Fourth, illegal wildlife trade can pose a threat to human health. Those involved in tackling the illegal wildlife trade have long pointed to the risks of zoonosis, the transmission of disease from (nonhuman) animals to humans. The biosecurity risks are clear: Ebola, Lassa, Marburg, and COVID-19 are zoonotic diseases that originated in wildlife (bats, chimpanzees, and other wildlife) and then "jumped species" to humans.[21] In early 2020 the World Health Organization (WHO) declared COVID-19 a global pandemic, and conservation NGOs quickly pointed to the origins of the disease in markets in China where live animals are traded and are in close proximity to humans. While the precise mechanism by which the virus jumped species (most likely from horseshoe bats) to humans is not yet known, the illegal and unsustainable trade in wildlife is the most likely original source of the disease. By June 2021 the pandemic resulted in 174 million cases and almost 4 million deaths worldwide.[22] In efforts to contain the spread of the virus several countries imposed significant restrictions, or "lockdowns," which saw economies grind to a halt as people were advised or legally required to minimize interaction to prevent further transmission. In April 2020 a group of two hundred conservation and animal welfare organizations published an open letter to the WHO calling for a permanent ban on all wildlife markets and a precautionary approach to the wildlife trade.[23] Other experts cautioned against calls for blanket bans, arguing that they could be inimical to the livelihoods of some of the world's poorest people and could have counterproductive conservation outcomes for some species that were sustainably traded.[24] For example, a statement from CITES clearly indicates that the impacts of the illegal wildlife trade include human and animal disease transmission, rural livelihoods, ecosystem health, biosecurity, and global security threats.

The cross-border smuggling of live animals and plants carries with it risks to human health through the spread of disease, some of which (such as the Ebola virus) are life threatening. Diseases, such as bird flu, can also be spread to food chains, leading to mass euthanasia of livestock herds. The introduction of alien species to habitats can ruin the natural biodiversity of countries or regions. The ease with which some wildlife contraband is smuggled across borders, often in significant quantities, demonstrates very real threats to national security and the biosecurity of states.[25]

In sum, the wildlife trade, in both legal and illegal forms, has important human security dimensions; these arguments resurface in the debates about how and why conservation is becoming more integrated with security. However, to understand the ways conservation and security are becoming integrated we must do more than rehearse the arguments about the human security dimensions of changing wildlife use and trafficking. Rather, in this book I examine the responses of the conservation sector to the sense of urgency produced by species losses, especially in the ways that they have sought to tackle the illegal wildlife trade. This focus is important, and while it is not the only response by the conservation sector, it has become an increasingly important and powerful means of approaching the question of how to save species. Therefore, the more traditional ways of thinking about security, as national and global security rather than as human security, is the core concern of this book.

SECURITY

The motives driving conservation and security can be very different, which makes the fact that they are being integrated in new ways all the more interesting. There are, of course, longer-standing historical collaborations, for example, the use of national parks for military training or hiring of former soldiers as park rangers (see chapters 3 and 7). The central, or urgent, conservation issue that needs to be addressed in the illegal wildlife trade is species declines and losses as a result of unsustainable use. Conservation is often referred to as a *crisis discipline,* which Soulé argues is a mixture of art and science, because it requires conservation biologists to act before knowing all the facts.[26] This is a common theme in the wide range of global environmental challenges, characterized by uncertainty, risk, incomplete information, and a need to predict possible consequences of decision-making. Environmental experts, including those in conservation, are asked to provide advice to governments, international organizations, the general public, NGOs, and businesses based on incomplete and uncertain information.[27] In the face of risk and uncertainty, environmental experts use the precautionary principle to try to avoid exacerbating the problem, that is, taking a precautionary approach in which policy makers must act on the basis of preventing potential harms (in this case species losses).[28]

By contrast, the aim of security in traditional terms is to prevent or tackle threats to global and national stability. Traditionally, security threats have been viewed through the lens of threats to states from other hostile states—most obviously expressed by tensions between the Soviet Union and the US, and their respective allies, during the Cold War. With the end of the Cold War came a growing concern about *new* security threats, including human and animal diseases, chemical and biological weapons, mass population displacements, religious differences, ethnic differences, underdevelopment, climate change, pollution, and shortages of land, food, and water.[29] This shift in thinking, away from threats to states from other states and toward more diffuse threats from a range of nonstate sources, was significant. The post–Cold War period also saw the rise in human security approaches. There was a marked rise in international interventions, faith in ideas of preemption and conflict prevention, use of force, and an increasing acceptance of private sector operators in delivering security.

Until now this has been confined to the human world, but the urgent need to save animals threatened by the illegal wildlife trade has produced another shift. Conservation is now a site of convergence between approaches to global security and saving the world's iconic species. Since the early 2000s, there has been a growing sense that wildlife must be secured not only to maintain ecosystems and biodiversity but also to enhance human well-being and secure global stability. In environmental debates, the environmental security perspective rose to prominence, which sought to establish a link between simple scarcities and violent conflict.

Briefly, debates in critical security studies, especially around securitization, explore how particular issues become discursively constructed as security challenges. The Copenhagen School, for example, argues that elite actors construct security issues by discursively framing them as such (referred to as *speech acts*). This then facilitates the development and implementation of new emergency measures, which are not just beyond the realm of normal politics but should be seen as a failure of normal politics.[30] The interventionist and security-oriented responses to the emergencies of the illegal wildlife trade often lie outside the realm of the normal politics.

Geographers have also engaged with the ways that the constructing and defining particular places by powerful actors can shape policies and interventions toward them; the literature on geopolitics and critical geopolitics is particularly useful here because it can help us understand the ways in which global-level debates about the illegal wildlife trade (its actors, trafficking routes, and beneficiaries) can ultimately shape and even determine policies, funding, and interventions to tackle it. In particular, Shannon O'Lear's research on environmental geopolitics and Simon Dalby's work on the intersections between environment and security cast light on how the framing of the illegal wildlife trade shapes policy design.[31]

While speech acts and ways of framing an issue are vitally important in redefining particular issues as security challenges, I am also concerned with the very real and everyday ways that security discourses and framings play out on the ground for people and wildlife. For example, Massé and Lunstrum's analysis of the Greater Lebombo Conservancy (GLC) between Mozambique and South Africa demonstrates how securitizing discourses and practices facilitated the enclosure of spaces for conservation, which then offered new opportunities for the benefit (financial and otherwise) of the private sector operator.[32]

In this book I examine the much wider context in which conservation initiatives to tackle the illegal wildlife trade take place. The aims of security and of conservation have clear differences, so their integration is not obvious. Therefore, it is important—and timely—that we carefully examine what happens when they are conjoined in this way.

FRAMINGS

The changes to conservation have been driven by the ways that the illegal wildlife trade is framed as a security threat. Indeed, the representation of environmental issues as a threat is a key part of shifting issues from the realm of normal politics to that of urgent problems that require exceptional kinds of responses.[33] Conservationists are well-positioned knowledge brokers in this field. They are powerful and persuasive voices because the impacts of the illegal wildlife trade are characterized by risk, uncertainty, urgency, and incomplete information. The framing of the illegal wildlife trade as a serious crime, organized crime, or source of threat finance is important because it has produced a shift in real-world responses to it. Through such framings, saving wildlife can be more easily integrated with national and global security strategies.

This shift in conservation has been facilitated by the argument that illegal wildlife trade constitutes a security threat. It has been achieved firstly by redefining and elevating the illegal wildlife trade from being a crime to being a form of *serious crime.* Secondly, by arguing that the trade can be a funding strategy for nonstate armed groups (militias, rebels, terrorist groups) and organized crime networks (including the Mafia, Russian Mafia, and Triads, among many others), Wyatt argues that the illegal wildlife trade is a growing area of interest for criminologists because of the value of the trade, its increasing visibility, and implications for national and global security.[34] Green criminology can be understood as the study of crime, harm, and injustice related to the environment.[35] It allows us to ask interesting questions around who or what constitutes the victim of an environmental crime. This is important in debates about the illegal wildlife trade. As Massé et al. argue, framing the illegal wildlife trade as serious and organized crime emphasizes the transnational aspects of illegal wildlife trade, its putative convergence with other types of "serious" crime, and its destabilizing potential. This then promotes and privileges responses, such as legal and judicial reform, criminal investigations, intelligence gathering, law enforcement technologies, and use of informant networks.[36] Therefore the integration of conservation with the notion of security risks encourage and then produce security-oriented responses on the ground.

For example, in 2015 the United Nations issued a resolution that identified wildlife crime as on a par with other forms of serious crime, including drug and people trafficking. In the resolution, the UN urged member states to increase efforts to tackle wildlife crime.[37] The reconfiguration of wildlife crime as a serious crime is echoed by a range of international institutions and conservation NGOs, including CITES, INTERPOL, the EU, World Wide Fund for Nature, United for Wildlife, Conservation International, the Nature Conservancy, and many others.

In February 2017, conservationists in NGOs and US government departments welcomed the fact that the illegal wildlife trade was specifically mentioned in Donald Trump's Executive Order on Enforcing Federal Law with Respect to Transnational Criminal Organizations and Preventing International Trafficking. The order states: "Transnational criminal organizations and subsidiary organizations, including transnational drug cartels, have spread throughout the Nation, threatening the safety of the United States and its citizens. These organizations derive revenue through widespread illegal conduct, including acts of violence and abuse that exhibit a wanton disregard for human life. They, for example, have been known to commit brutal murders, rapes, and other barbaric acts."[38]

The executive order identifies wildlife trafficking as a specific category of organized and transnational criminal activity and therefore as a threat to US national security. It places wildlife trafficking on a par with other serious threats to national security, which is indicative of how wildlife trafficking has become a much higher priority than in the past. This shift toward the idea that the illegal wildlife trade is a serious crime has been, in part, facilitated by the ways that wildlife trafficking and poaching are increasingly subsumed under the banner of security threats (discussed more fully in chapter 2). This framing has contributed to the sense that there is a need to take urgent action, which can include the use of force.

If we examine specific cases, they reveal the ways that debates about the illegal wildlife trade often revolve round stark definitions of rangers as heroes and poachers as cruel criminals. Such characterizations rely on drawing a moral boundary between simplistic caricatures of good guys versus bad guys, which is evident in several conservation NGO campaigns about poaching of large charismatic species, particularly elephants and rhinos. In 2017 the Wildlife Conservation Society (WCS) 96 Elephants Campaign was organized around three pillars: Humans and Elephants, Terror and Ivory, and Heroes and Hope.[39] By 2020 the campaign had shifted to pointing to its success: gaining an ivory ban in the US, the development of the Eliminate Neutralize Disrupt (END) Wildlife Trafficking Act, and the two million other actions the WCS and supporters had taken to save elephants in Africa. The campaign is now organized around "Stop the Killing, Stop the Trafficking, Stop the Demand."[40] But the 2017 campaign is illustrative of how particular narratives were mobilized to gain wider support. The final pillar of the 2017 campaign, Heroes + Hope, illustrates the ways rangers can be cast as heroes who have names, stories, and identities, whereas poachers appear only as pervasive but depersonalized enemy, without names or further detail about who they are. The campaign points to the "unflinching courage" of rangers: "Often, what stands between life and death for an elephant are rangers. These brave men and women know the land, the people and their hardships, and are the eyes and ears for other conservationists. With their frequent patrol efforts, they put their lives on the line to protect wildlife, and to ensure that conservation efforts work."[41]

The rangers as the central heroes of the campaign are supported by a wider range of individuals cast as boots on the ground, including conservation biologists, community leaders, and politicians. One ex-poacher does appear in the list of heroes: Thomson Tembo, described as a "once notorious ivory poacher" in Zambia, who has now begun working with WCS to develop community projects and train others in sustainable agriculture.

The process of moral boundary drawing between poachers and rangers relies on the idea that conservationists are engaged in a justifiable, even just, war to save wildlife.[42] Poachers are defined as combatants and legitimate targets, in contrast to the civilians described as the people who suffer hardships. Poachers are presented under the theme of Terror + Ivory, which intersects with high-profile international claims that ivory poaching and smuggling is a source of funding for global terrorist networks. Indeed, the 96 Elephants Campaign information makes direct reference to this, referring to ivory as the "white gold of jihad";[43] yet this characterization of poached ivory has been heavily criticized as lacking in credible evidence (see chapter 4).[44] In the 96 Elephants Campaign, WCS clearly states that poachers are not motivated by poverty but instead are part of larger criminal and terrorist networks. This in itself sets up the poacher as interested in material gain, as a terrorist or criminal who needs to be tackled forcefully via strategies informed by military and/or criminological approaches. For example, the campaign website reads: "How They Kill: Brutal and sophisticated. Both words describe the armed militants poaching Africa's elephants. The killers use helicopters, GPS equipment, night-vision goggles, and automatic weapons to find and mow down elephants, then hack their tusks out with an axe—an atrocity often committed while the animal is still alive."[45]

In this campaign, conservation is redefined as an urgent security issue through the language used to present poachers and rangers. The campaign language also reflects the arguments about the need to intervene in complex emergency situations, especially when they are linked to claims that underdevelopment encourages violent conflict and insecurity. Duffield refers to this as the "security-development nexus," which sees underdevelopment as inherently dangerous.[46] The campaign by WCS is an example of how wording and imagery can directly and powerfully shape the ways in which we understand the illegal wildlife trade, the drivers of poaching, and the roles of rangers as wildlife defenders. I do not mean to suggest that these presentations are untrue, rather that the ways we talk about illegal wildlife trade can have material power on the ground—it shapes funding priorities, the design of policies, and the ways that communities living near national parks are treated by law enforcement.

Academic researchers, especially those from the field of political ecology, have been concerned with highlighting the relationships between conservation, violence, and conflict. They highlight the forms of violence produced by state-led initiatives to control natural resources and the benefits that accrue from establishing state control.[47] Political ecologists have always been concerned with understanding who benefits, who loses and why, in the changing relations between humans and the environment. Neumann, for example, argues that war is a common model for biodiversity protection in Africa, where protected areas can be spaces of violence, underwritten by a Malthusian fear of the poor and their claims on resources.[48] Therefore, conservation can be both a *means* of and a *reason* for violence.[49]

Current shifts in conservation mean these important debates need further development. By applying a political ecology lens to the ways that conservation and security are becoming integrated in responses to the illegal wildlife trade, this book is intended to do just that. An important part of the story is the ways that the security phase of conservation builds on the earlier, more market-oriented and business-friendly phase of conservation.[50] This relied on a process of enclosure, of drawing boundaries around landscapes, forests, watersheds, and wildlife to make them amenable to and tradable on global markets. This practice is often referred to as "selling nature to save it."[51] The environment has become framed as a security issue, in part precisely because wildlife, forests, marine resources, and so on have all become financially valuable, in carbon markets and wildlife tourism, for example.[52]

Market-oriented approaches are also discernible in the security phase of conservation. It is apparent in the arrival of new private security companies offering anti-poaching training or in technology firms developing new products for use in wildlife surveillance and monitoring (see chapters 6 and 7). Furthermore, approaching the illegal wildlife trade as a security issue has opened opportunities for new players to enter into conservation, enabling them to tap into the enhanced levels of funding available and to develop new markets for security-related products and services.

PRACTICES

The ways that the illegal wildlife trade is defined and presented as a global security threat has material impacts on the ground. These range from the growth of surveillance technology in anti-poaching operations to weapons training for rangers. Militarization is the most obvious and "spectacular" manifestation of the integration of conservation and security around the illegal wildlife trade (see also chapters 3 and 7), but it is only one expression of it.

Political ecologists have been at the forefront of debates about the growing intersections between conservation and militarization, which is variously referred to as green security, green militarization, green violence, and green wars.[53] A rollout of more militarized responses is justified via appeals to the urgent need to prevent the loss of a significant form of natural heritage and important species for the whole world.[54] As Verweijen suggests, while this work points to the importance of understanding the links between the broader structural context and the operation of militarized forms of conservation, it does not trace how specific acts of violence are committed. By focusing on the micro-dynamics of violence in conservation in Virunga National Park, Verweijen seeks to map the "kill chain" and understand the ways that specific acts of violence are produced and uncover the (often obscured) factors that contribute to them.

Direct forms of violence by rangers against civilians living near Virunga National Park include intimidation, beatings, rapes, torture, and killings. Verweijen's informants expressed their fears about entering the park. They recounted being beaten with rifle butts, with women being stripped before beatings, and at least one case of the repeated rape of a minor. Experiences like these result from a range of transnational regional, national, and subnational factors. Conservation in Virunga is highly transnationalized and is characterized by considerable power asymmetries. Specific instances of violence can be produced by a complex mix of group dynamics, military-style training by Belgian commandos, support for greater enforcement from funders and donors, inadequate training in human rights, and conflict de-escalation, among many other factors.[55]

More militarized forms of conservation are being used in a wide range of conservation initiatives—not just in the arena of the illegal wildlife trade. For example, in Nigeria militarized strategies, including armed rangers, are used to protect forests designated for global climate change mitigation schemes, such as the UN Reduced Emissions from Deforestation and Forest Degradation+ (UNREDD+). Greater emphasis on enforcement and the justification of use of force has been articulated as necessary to protect standing forests that have a new economic value on international carbon offset markets and to address global climate change. Local communities infringing on the forest are seen as the key threat.[56] While sub-Saharan Africa may offer the clearest examples, similar dynamics are found elsewhere, for example, the rhino protection in Kaziranga National Park in India, protection of ecotourism sites in Colombia and Honduras, forest conservation in Guatemala, and securing REDD+ forests in Laos.[57] These examples indicate that the security turn in conservation goes well beyond the illegal wildlife trade, and it is not confined to Africa either.

FUNDING SOURCES

The need for funding is a core theme in debates about conservation, and conservation NGOs have repeatedly argued that there is a need for a much greater level of funding. For example, a 2020 report from the Nature Conservancy estimated that in 2019 spending on biodiversity conservation was between US$124 and US$143 billion per year; this includes financing from donors, biodiversity offsets, nature-based solutions and carbon markets, philanthropy and conservation NGOs, with the largest percentage (57 percent) coming from domestic state budgets and taxes. The report argues that this falls far short of the needs of conservation, estimated as between US$722 and US$967 billion per year.[58] However a report for OECD provided very different estimates: US$78–91 billion per year (2015–2017 average), comprised of public domestic expenditure (US$67.8 billion per year), international public expenditure (US$3.9–9.3 billion per year), and private expenditure on biodiversity (US$6.6–13.6 billion per year).[59]

The intense interest in illegal wildlife trade means that more funding has been made available for anti-poaching, demand reduction, and anti-trafficking. The European Commission alone established a EUR 350 million (approximately US$380 million) trust fund to promote conservation in Africa. Regional organizations, including the African Union, are also developing their own strategy documents for halting poaching.[60]

Since 2015, the largest single funder of initiatives aimed at tackling illegal wildlife trade has been the Global Environmental Facility (GEF) Global Wildlife Program. It is a World Bank–led initiative, which began with a budget of US$131 million, rising to US$187 million by 2020 (through financial contributions from the UK government), with an aim to raise US$704 million in total.[61] According to a senior World Bank official, the origins of the Global Wildlife Program lie in discussions about tiger conservation in 2007. By the time the Global Tiger Forum was held in St. Petersburg in 2010, the bank, along with other high-profile partners, were developing a global strategy to tackle the illegal wildlife trade. Since its inception the projects and initiatives of the Global Wildlife Program have been informed by the key concerns of partner governments in Asia and Africa.[62]

These figures should be treated with caution—it is very difficult to estimate the total spending on biodiversity conservation globally as different reports rely on a range of methods for calculating total amounts, and it can often be impossible to tease out the precise figures. For example, USAID as a key donor may provide funding to WWF-International, which then acts as a donor to local level conservation organizations or national chapters of WWF using a mixture of USAID funding and its own revenue. As a result determining exactly how much funding is available for which activities can be very challenging, and there is capacity for double counting.

Donors and philanthropic foundations play an important role in shaping and supporting efforts to tackle the illegal wildlife trade. A review by the Global Wildlife Program of the World Bank GEF in 2016 found that a total of US$1.3 billion was committed by twenty-four international donors between 2010 and June 2016.

This provided funding for 1,105 projects in sixty different countries, as well as regional and global projects. Of the donors, the top five were the GEF, Germany, the United States, the European Commission, and the World Bank Group, which together contributed US$1.1 billion of the total funding (86 percent). Sixty-three percent of the funds went toward efforts in Africa (US$833 million), 29 percent to Asia (US$381 million), 6 percent to global programs and initiatives (US$81 million), and 2 percent to projects covering both Africa and Asia (US$35 million). The top five recipient countries were Tanzania (8 percent), the Democratic Republic of Congo (5 percent), Mozambique (5 percent), Gabon (3 percent), and Bangladesh (3 percent). Forty-six percent of the funding supported protected area management, while 19 percent went to law enforcement including intelligence-led operations and transnational coordination, 15 percent for sustainable use and alternative livelihoods, 8 percent for policy and legislation, 6 percent for research and assessment, and 6 percent for communication and awareness raising.[63]

These numbers reveal an interesting pattern of flows of money toward particular beneficiaries. The largest chunk of funding was allocated to projects in Africa (67 percent), and 10 percent went to projects that linked Africa and Asia; national governments were the main beneficiaries of this funding rather than NGOs, private sector, research groups, or multilateral organizations.[64] In 2018–19 the GEF was the biggest single donor providing funds for projects to tackle the illegal wildlife trade. In 2018 the GEF announced it had committed a further US$168 million over its new funding cycle (the GEF-7) from 2018 to 2022, an increase on the US$131 million it had committed in 2015–18.[65] In the first round an additional US$704 million was mobilized in co-financing from governments, donors, private foundations, private sector, and civil society.[66]

Funding from US Fish and Wildlife Service (USFWS) and the UK's Department for Environment, Food and Rural Affairs (DEFRA) also reveals similar patterns. According to Massé and Margulies, between 2002 and 2018, the USFWS Division of International Affairs provided assistance to 4,142 projects

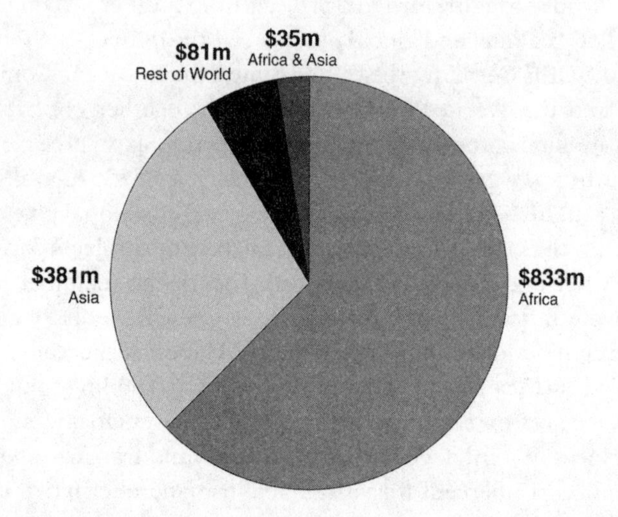

Amount of funding received from international donors to tackle illegal wildlife trade by geographical location (GEF figures).

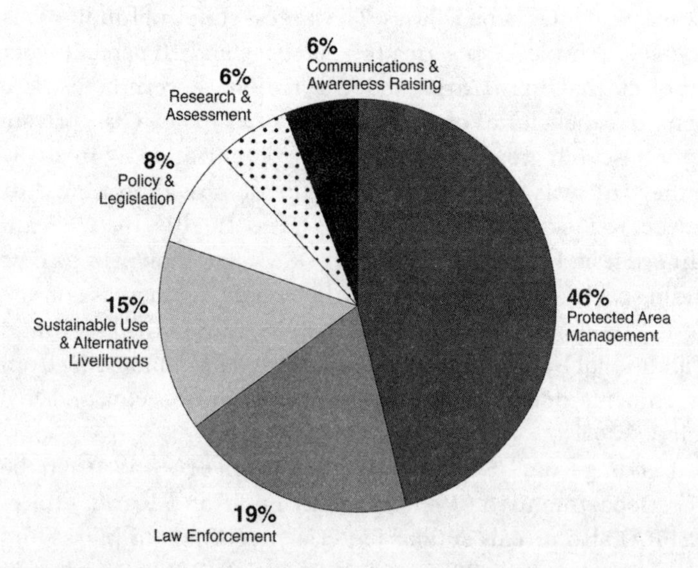

Percentage of funding received by project type (GEF figures).

across 106 countries worth over US$301 million. They suggest that the illegal wildlife trade went from being a boutique issue to one of central importance, and as a result an increasing portion of foreign assistance for biodiversity conservation was allocated to projects that centered on illegal wildlife trade. In 2014 the US Congress, and specifically the Sub-Committee on Department of State, Foreign Operations, and Related Programs, allocated US$45 million in the foreign assistance biodiversity budget to tackle wildlife trafficking, an amount increased to US$55 million in 2015, US$80 million in 2016, and almost US$91 million in each of 2017, 2018, and 2019, at the expense of other conservation priorities.[67]

Similarly, the UK government has made tackling the illegal wildlife trade a policy priority. The Illegal Wildlife Trade Challenge Fund was established in 2013, and by 2019 it had allocated just over £23 million to seventy-five projects. It is notable that a few conservation NGOs stand out as key beneficiaries of the fund. WCS received a total of £5,032,784 for sixteen projects; the next largest beneficiaries were Zoological Society of London (ZSL) with seven projects totaling £3,210,400, and Flora and Fauna International (FFI), with five projects totaling £1,768,492.[68] Of course this is partly a reflection of the numbers of applications they submitted and the ability of their grant writers to tailor the applications to the relevant criteria.

Further, it is interesting to note the relative balance of projects spread across three themes: developing sustainable livelihoods for communities affected by illegal wildlife trade, 6 funded projects; strengthening law enforcement and the role of the criminal justice system, 62 funded projects; and reducing demand for wildlife products, 7 funded projects.

The disparity is clear: approximately ten times more funding went toward projects associated with law enforcement and the criminal justice system than to the other priority areas. These results reflect two things. First, there were more applications that identified law enforcement as their lead theme; second, the criteria required applicants to make a clear link between the project and potential impacts during the life of the project via a detailed

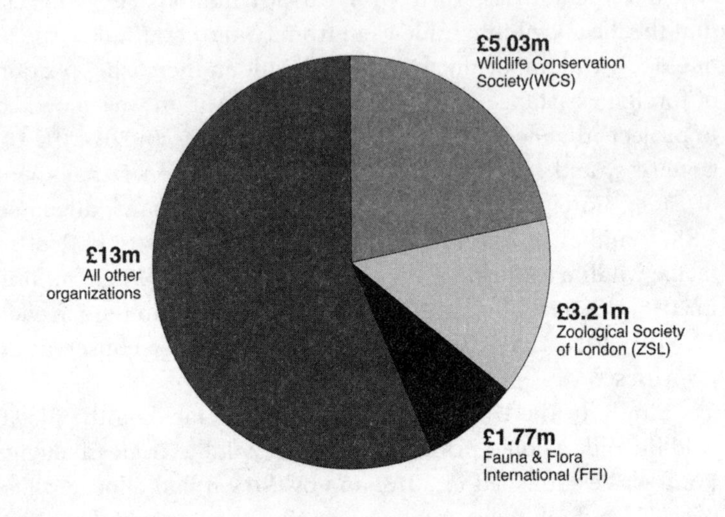

DEFRA funding received by type of organization (£M).

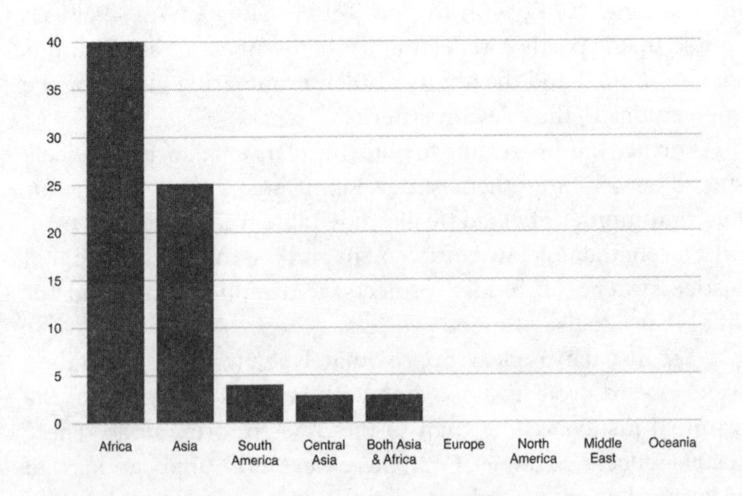

DEFRA number of projects funded by geographical location.

"logframe." It takes much longer to demonstrate success in terms of demand reduction and the development of sustainable livelihoods.[69] The geographical breakdown of funded projects also indicates there were more projects in certain regions: Africa (40), Asia (25), South America (4), Europe (0), North America (0), Middle East (0), Central Asia (3), Oceania (0), mixture of Asia and Africa (3).[70]

These figures from GEF, USFWS, and DEFRA serve as illustrations of the profile of funding available for conservation activities focused on tackling the illegal wildlife trade. The increases in available funding have facilitated the shift toward a more security-oriented approach in conservation efforts aimed at poaching and trafficking. But is this a blunt relationship between more funding and a sudden shift to security-oriented practices?

One intelligence specialist noted that donors "are immensely sensitive to get engaged in anything with a security side, it has risks."[71] Donors like USAID, the GEF, and the UK government have been careful about what kinds of initiatives they will get involved in, but current responses to the illegal wildlife trade do raise concerns about whether funding is being used in ways that actually breach human rights (see chapter 3). As one intelligence specialist said: "We are getting involved with donors who are keen to help but don't know how to help. To save time we are joining up the protected parks so they can articulate the problem. Because the world of information collaboration is an unfamiliar one for most people it just looks complicated and full of difficulties."[72]

The involvement of philanthropists also has the capacity to reshape conservation in ways that fit the vision of specific wealthy individuals. In September 2017 I ran a knowledge exchange workshop for conservation professionals to discuss what they saw as the patterns, challenges, and problems arising from the integration of security concerns with conservation. I had assumed that the group of twenty-five would be generally favor this trend, but instead they collectively expressed concerns and frustrations. One theme came through strongly: while the funding from individual philanthropists was welcome, these people often wanted to see quick results for their investments. On the ground, in some instances, it was producing pressure to secure arrests of suspected poachers, leading to more willingness to resort to firearms, with more injuries for suspects and wildlife officials alike. At the more extreme end, there were concerns that there might be more pressure to engage in shoot-to-kill.

Almost a year later when I ran another focus group with a different set of conservation professionals, one pointed out that they were in an impossible position: how could they not partner with security companies and philanthropists in the circumstances? They were facing difficult situations: staff worked in areas where they were aware that members of local communities were subject to punishment beheadings by armed groups, and they feared for their own safety.

Philanthropists, often with the best intentions, can have a disproportionate influence on the direction of conservation policy.[73] Since philanthropists have the available financial means, they are essentially able to purchase the kind of conservation vision they favor.[74] We need a better understanding of these dynamics—why they have developed, how they are changing approaches to the illegal wildlife trade, and what the future implications are for both security and for conservation.

It is reasonable to ask why any of this should concern us. Is there a case for seeing the illegal wildlife trade as a security issue if it threatens species with extinction and contributes to national and global insecurity? Certainly, the argument that the illegal wildlife trade is a security issue has taken center stage in international policy networks. In this book I examine the different ways that the illegal wildlife trade is debated on an international scale, the politics of presenting rangers and poachers in particular ways, the role of the political economy of the private security industry, and the long history of engagements between militaries and the conservation sector. It is important to render the wider context of the illegal wildlife trade more visible. There are multiple steps that lead to the point in time when an elephant is poached, pangolin scales are consumed, an orchid is removed from its habitat, or a parrot is caged, to end its life as a luxury pet. Rather than focusing on these key moments, harrowing and challenging as they are, here I zoom out to take a more expansive view, which addresses the wider politics of the response of the conservation sector to the illegal wildlife trade and its effects on people and wildlife.

Current approaches to understanding the illegal wildlife trade and the threat it poses to global stability may appear, on the surface, as an ideal example of a *new security threat,* which neatly fits with established and mainstream environmental security thinking. However, this does not capture the richness, depth, and significance of the changes in conservation that have resulted from its integration with security. By focusing on responses to the illegal wildlife trade it is possible to trace and analyze the differing aims of conservation and security, the importance of how it is framed as a security issue, and how that leads to particular material outcomes on the ground.

This book explores further how the convergence of conservation and security is changing what conservation is on the ground. In areas where illegal traded species are found, it is defining what conservation *can* be and, more importantly, *cannot* be in the future. As the security-oriented approach opens up new partnerships with private military companies, national armies, intelligence services, and tech companies, it is important to reflect on how these shape interactions with other potential collaborators, including local communities, environmental activists, and grassroots development NGOs.

Taking a security approach to illegal wildlife trade will have limited success and will not provide a long-term workable solution to biodiversity losses. It may save a few plants or animals in certain locations; it may even prevent an extinction or two. All of this is important and should not be overlooked, but the underlying pattern of species losses will continue because security-oriented approaches do not address the underlying drivers of extinctions and reduced numbers of wild animals and plants. They sidestep the ways that dynamics of global wealth and inequality, produced by the wider economic system, can drive the trade. Wealth can drive demand, yet poorer communities will often end up as the suppliers at the other end of the chain, taking risks, receiving a fraction of the value, or living with the more hostile and forceful approaches of conservationists, states, NGOs, and a private sector seeking to protect wildlife. Further, the security-oriented approach actually opens new opportunities for those seeking new markets and profits, thereby deepening and driving the very logics that produce the problem in the first place. In the chapters that follow I explore these themes further.

2 FRAMINGS MATTER

The Illegal Wildlife Trade as a Security Threat

The illegal wildlife trade has become a high-profile global issue. This is partly explained by concerns about rising rates of poaching, but it is also as a result of the redefinition of illegal wildlife trade as a global security issue. Illegal wildlife trade and wildlife crime are not necessarily the same thing, but they are often used interchangeably or fused together in debates about how the illegal wildlife trade constitutes a security threat. Does it matter how illegal wildlife trade is defined? How do we decide what *is* and what *is not* a wildlife crime? And in what ways do key actors draw the links between illegal wildlife trade, wildlife crime, and security issues? CITES, for example, states that wildlife crime shares many of the characteristics of other types of transnational crime, such as the trade in narcotics; rather, "wildlife law enforcement officers also often lack parity with their counterparts in Customs and Police services, and due to the diverse nature of wildlife crime, may be ill-prepared to effectively respond to the challenges it presents."[1] This view has already begun to change, and what implications will that have for both conservation and security, as well as for wildlife and people?

Before 2008, when there was a rise in poaching and trafficking, conservation NGOs, law enforcement agencies, and international organizations working on the illegal wildlife trade struggled to gain the attention of governments. There were exceptions, such as the global-scale moves toward banning the ivory trade at CITES 1989 and the development of the International Tiger Conservation Forum, hosted by the Russian Federation in St. Petersburg in 2010. Although poaching of elephants, rhinos, and tigers gained some international attention, it was not presented as a threat to global security but instead as an environmental issue. As Lorraine Elliott points out, transnational environmental crimes are often not taken seriously within the broader policy and enforcement community, and they are perceived as a low risk, high reward activity for organized crime networks.[2] This is changing. Wildlife-related problems were once regarded as boutique issues, the responsibility and concern of the conservation community, but the widening range of impacts across the world have generated a greater degree of interest from a wider variety of sectors.[3]

How does the framing of the illegal wildlife trade as a wildlife crime matter for conservation practice on the ground? Articulating the illegal wildlife trade as a wildlife crime and as a security threat can obscure the deeper dynamics that drive the trade in the first place. These dynamics include the global structures of inequality, often coupled with deeply held historical and cultural attitudes toward wildlife products. Wildlife products are in demand by wealthier communities around the world yet are often sourced from areas inhabited by the world's poorest and most marginalized communities, making the inequities more glaring. What are the challenges in defining a wildlife crime? How has trafficking in wildlife been presented as an arena of organized crime networks? And how do these intersect with arguments about other illegal trades? These are some of the questions I address in this chapter.

WHAT IS A WILDLIFE CRIME?

Wildlife is traded at a global scale, both legally and illegally. TRAFFIC estimates the legal trade of wildlife products into the EU alone is worth nearly EUR 100 billion.[4] The scale of illegal trade is more difficult to estimate because of its clandestine nature. The EU estimates that the global illegal wildlife trade is worth between

EUR 8 billion and EUR 20 billion annually.[5] The range of estimates from different agencies value it between US$7 billion and 23 billion annually.[6] Valuing the trade is challenging because it is difficult to separate out legal and illegal trades in particular species because they are often deeply intertwined and profiles of demand and supply can change very rapidly.[7]

There is no universally agreed definition of wildlife crime. As the UN Office on Drugs and Crime (UNODC) points out, there are international legal instruments which set out the definitions of trafficking in humans, drugs, and weapons, but no such legal definitions exist for trafficking wildlife. It is even more challenging because different countries have different regulations that govern the uses and conservation of different species.[8] The debates about illegal wildlife trade and about wildlife crime involve a confluence of different approaches. The conservation community has always had a concern about the impact of the illegal wildlife trade on species, but the more recent articulation, or branding, of it as wildlife crime and as a form of serious crime also facilitates raising security concerns. The law enforcement and security communities have long regarded illegal wildlife trade as a crime because it is an illegal activity, but it has not been a high priority. Donors, national governments, and international organizations have begun to elevate illegal wildlife trade from the status of crime to serious crime or organized crime, which points to its intersections with security concerns. The distinction between crime and security is often lost in debates about the illegal wildlife trade. It must be remembered that *criminalizing* and *securitizing* tend to involve two distinct processes. Traditionally *securitization* centers on framing an issue as above politics and requiring emergency action, including extralegal options. In contrast, *criminalization* usually entails policing, law enforcement, and legal instruments and remains within the realm of normal politics. The debates about the illegal wildlife trade fuse both criminalization and securitization to present wildlife crime as a global security threat, prosecuted by organized criminal networks, armed groups and even terrorist networks. This argument is then deployed to justify the use of emergency, and often extra-legal and violent, measures, which eventually become part of the realm of normal politics, of which conservation practice is a part.

Gore argues that the rising interest from global policy makers results from a growing agreement that wildlife losses have produced a convergence of threats to ecosystems, geopolitical stability, national security, human health and well-being, and future generations.[9] The arguments around conservation-crime convergence is characteristic of a new security turn in conservation more broadly; this promotes and privileges responses anchored in legal and judicial reform, criminal investigations, intelligence gathering, law enforcement technologies, and informant networks.[10] It has also produced a great deal of uncertainty among policy makers about what sorts of data are needed and available to inform effective solutions.[11] Indeed, this was echoed by a representative from the then UK Foreign and Commonwealth Office, who noted that there was a dearth of information yet a need to *look for evidence* of wildlife trafficking as a security threat (my emphasis). The representative described how the office was operating to persuade other governments to take the issue of illegal wildlife trade seriously: with some governments, they might not be interested in it per se, but were receptive to discussing the need to tackle financial crime, of which wildlife trafficking could be a part.[12]

It is clear that referring to the illegal wildlife trade as wildlife crime has gained traction in debates about conservation. The term *wildlife crime* is used widely, but it is very imprecise and fuzzy and is often linked to a range of other issues. Wildlife crime is not the same as illegal wildlife trade, which is a narrow set of activities in breach of national level or CITES regulations. Wildlife crime, by contrast, is much broader. For example, in the UK, the National Wildlife Crime Unit deals with CITES issues and the illegal wildlife trade, but it also tackles a much wider range of activities, such as raptor persecution to protect game birds, badger persecution, deer and fish poaching, hare coursing, bat persecution, and disturbance or destruction of freshwater pearls.[13]

There are two particular ways the term "wildlife crime" is used to denote a security threat. First, it is used in the sense that wildlife crime is a serious crime and constitutes a security threat to governments. Second, it is increasingly used as part of a shift in language away from militarization (which has attracted criticism) and toward law enforcement. Although there has been a shift from militarization to the language of wildlife crime, debates about the illegal wildlife trade still rely on somewhat blunt characterizations of the trade and the people involved in it. As one conservationist based in Washington, D.C., put it, "Right now we are dealing with the cartoon version, security, organized crime caricatures, it's way more complicated than that."[14] Despite these complexities, defining the illegal wildlife trade as wildlife crime that constitutes a global security threat is regarded in some policy circles as the most effective for gaining attention and generating funding and policy commitments. A senior representative of the UK Foreign and Commonwealth Office underlined this, arguing: "Breaking down what we mean by security threat will be helpful in helping us tackle IWT. I am absolutely convinced it is the right way to talk about it, it is really helpful, but security means different things to different people."[15] This statement reveals the complexity of thinking about illegal wildlife trade as a security issue to gain attention at the international level.

Comparing the definitions from three conservation actors—an international convention (CITES), a US Government Department (USFWS), and a conservation NGO (WWF-International) — reveals the differences in how wildlife crime is defined, which ultimately shapes policy responses to it.

First, CITES defines wildlife as all flora and fauna, and it takes its definition of crime from ICCWC: "It refers to acts committed contrary to national laws and regulations intended to protect natural resources and to administer their management and use. . . . It may also include subsequent acts, such as the processing of fauna and flora into products, their transportation, offer for sale, sale, possession, etc. It also includes the concealment and laundering of the financial benefits made out of these crimes."[16] For CITES, wildlife crime encompasses the illegal trade in flora and fauna in its broadest possible sense, including transport, possession, and laundering of financial benefits realized from the trade.

The rising interest in poaching and trafficking as a wildlife crime was partly the result of dedicated and sustained activity by John E. Scanlon when he was secretary-general of CITES, from 2010 to 2018. Looking back on his time as secretary-general, Scanlon regarded the fact that the world rediscovered CITES as one of the successes of his leadership period. He explained this as the result of the rises in poaching and because of his own active involvement in persuading key governments and international organizations that the illegal wildlife trade should be seen through the lens of wildlife crime as a serious organized crime.[17] His leadership drew attention to the illegal wildlife trade and to CITES's role in tackling it. It did so in very particular ways, cementing the idea that the illegal wildlife trade needed to be tackled as a matter of crime and security, rather than primarily of conservation, inequality, or development.

The second example is from USFWS, which defines a wildlife crime as "activities which include smuggling, or trafficking, the poaching or other taking of protected or managed species and the illegal trade in wildlife and their related parts and products; unlawful commercial exploitation; habitat destruction; poisoning incidents."[18] This definition has some points in common with that of CITES around illegal wildlife trade, smuggling, and trafficking; the key difference is the addition of habitat destruction as a form of wildlife crime. It raises a different set of questions about how we identify the perpetrators of wildlife crime. The introduction of habitat destruction opens up the possibility of identifying governments and corporations as perpetrators of wildlife crime, in addition to individual poachers and trafficking networks. This point is crucial in determining how we think about wildlife crime and respond to it.

The third example of defining wildlife crime comes from the world of conservation NGOs. WWF-International interprets wildlife crime through the lens of the illegal wildlife trade and its impacts on wildlife populations. The joint Wildlife Crime Initiative between WWF-International and TRAFFIC International was launched in 2014 in order to address the global poaching crisis. It focuses on four pillars: "stop the poaching, stop the trafficking, stop the buying, and international policy."[19] This framing very much focuses on individuals and networks of poachers and traffickers and not on the wider processes of habitat destruction included by USFWS.

It is clear from these brief examples that three different kinds of actors in the debates about the illegal wildlife trade have slightly different interpretations of what constitutes a wildlife crime. Furthermore, they have distinct approaches as to who or what might be responsible for perpetrating wildlife crimes, which shapes their policy responses.

WILDLIFE CRIME AS THE MUNDANE AND EVERYDAY

In debates about the illegal wildlife trade high-profile charismatic species such as elephants, tigers, and rhinos get the greatest attention. However, the trade covers a whole range of less well-known species and can in fact be very ordinary and mundane. Put simply, the illegal wildlife trade can be *un*organized and *un*charismatic.[20] When the focus is on larger and more charismatic species, other aspects of the trade can get little attention. This thought was echoed by a US Government official who pointed out that other trades tend to be overlooked, for example, those in rosewood, amphibians, reptiles, birds, seahorses, and animals destined for the pet trade.[21] The same argument was made by a member of the European Parliament (MEP) who was actively involved in raising the profile of the illegal wildlife trade as an issue for the EU. This MEP suggested that "it is a lot easier to grab people's attention when you talk about elephant poaching, but if you go around talking about pangolins that are probably far more traded and in far more trouble, but maybe people don't know what a pangolin is."[22]

The illegal wildlife trade throughout Europe and in European species is also overlooked compared to higher profile and more charismatic African and Asian animals. One senior European law enforcement official put it this way: "Public attention on wildlife crime is very focused on ivory trafficking, rhino horn. What is not taken into consideration is that in Europe indigenous species are also being traded but they don't get that kind of attention. In Scandinavian countries, there are birds of prey harvested and traded into Arab countries. Or amphibians from Germany, which are taken and traded. Songbirds are also something where there is an illegal trade in EU countries."[23]

The European Union is a key site for illegal wildlife trade: it is a transit point in the trade, and it is both a source of wildlife products and a consumer of them. The trade in European eels is an example that is often overlooked and poorly understood. Since 2010 trade in wild-caught European eels from the EU has been banned. However, juvenile eels, also known as glass eels, are still being taken from rivers and exported to Asia and farmed, before being re-exported or consumed locally when they are adults. For example, the European Union Agency for Law Enforcement Cooperation (EUROPOL) supported Operation Elvers, which tracked the illegal trade in glass eels to farms in China from Spain and Portugal, with support from collaborators in Morocco.[24] The Sustainable Eel Group suggests that 16 percent of European eels are illegally captured and trafficked, while 7 percent are legally caught but then illegally exported.[25] This indicates that there is a complex interrelationship between legal and illegal activities.[26] Examining the illegal wildlife trade in the EU breaks down the assumption that there are regions or places that consume wildlife (often identified as Asia, specifically China) and places that are the sources of wildlife products (often assumed to be Africa).

The EU Trade in Wildlife Information eXchange (EU-TWIX) database shows that there are four important trade routes into the EU: large mammals (elephants, rhinos, and big cats) from Africa and South America transiting through major hubs in Europe for export to Asia; coastal smuggling of leeches, caviar, and other fish, as well as reptiles and parrots for the European pet trade; endangered birds traded from southeastern Europe to southern Europe; and finally Russian and Asian wildlife traded into Europe via eastern European land routes.[27] EUROPOL has also identified France, Belgium, the UK, the Netherlands, and Germany as the more commonly used wildlife trafficking transit hubs, especially airports and ports.[28] Heathrow in the UK is a major entry and transit point for wildlife;[29] Paris's Charles de Gaulle Airport is an important hub for wild-caught meat (popularly referred to as bushmeat) from Africa;[30] and smaller European airports with direct connections to Africa and Asia are emerging as new trade hubs.[31] The central role of European transport and trade hubs in the legal and illegal wildlife trades is often overlooked. Furthermore, the trades in European species do not elicit the same intense interest as the high-profile, charismatic African and Asian animals. Nevertheless, they are really important parts of the overall story.

The ways that less charismatic species get overlooked is borne out in debates at the biennial Conferences of Parties (CoP) for CITES. In among all the high-profile debate about ivory and rhino horn trade at CITES CoP17 in Johannesburg in 2016, many other species were discussed and newly listed under CITES. These species did not attract a great deal of public, media, or NGO attention. For example, the EU and Vietnam jointly tabled a motion to list the Vietnamese psychedelic rock gecko on CITES Appendix 1. The lizard was under threat due to a growing demand from the pet trade, centered in the EU and the Russian Federation. Live pairs, taken from the wild in Vietnam, were being offered for approximately EUR 2500–3500 at reptile shows.[32] The decision at the CITES CoP17 to list the species as CITES Appendix 1, a listing which translates into a total ban on international trade in the lizards, was not as widely debated or reported as the decisions on trophy hunting, ivory trade, and attempts to reopen a legal trade in rhino horn.

The illegal trade in wild plants is even more overlooked. The plant community refer to this as *plant blindness* in wildlife trade debates.[33] Approximately 5,800 species of animals and 30,000 species of plants are listed by CITES.[34] Of those 30,000 listed plant species, 70 percent are orchids.[35] In the debate about illegal wildlife trade it is important to remember that it is an extensive global trade, and also that often it includes everyday, mundane products that many of us do not even think of in connection to wildlife trafficking.

Dudleya farinosa. Photo credit: Jared Margulies.

The recent expansion in illegal trade in *Dudleya farinosa* illustrates the complexities and contradictions in debates about the illegal wildlife trade. *Dudleya farinosa* is a succulent plant species found along the west coast of the United States; it is not CITES listed because it is not considered under threat from the trade.

Legal mechanisms exist for its trade, so as Margulies points out, it is surprising that an illegal trade exists at all. It is defined as illegally traded if it is taken from private or public lands without permission, or if it sold or exported without the required permits. Illegal wild collection of the plant expanded from 2017 due to demand by collectors, which did attract some media attention but not to the same degree as rhino and elephant poaching.[36] The story of the collection of this plant further reveals the intersections between legal and illegal trade. Examining the plant trade or the role of Europe as a place for sourcing, transiting, and consuming wildlife rapidly breaks down assumptions about how the trade actually operates.

THE BOUNDARIES OF WILDLIFE CRIME?

It is important to understand more fully the different ways that the term "wildlife crime" is deployed, by whom, and with what kinds of effects. The focus on organized crime misses the ways that legitimate businesses can be engaged in illegal wildlife trade at different points in the supply chain. Daan Van Uhm calls these "green collar crimes," defined as environmental crimes committed by legally registered companies involved in illegal activities or which use their infrastructure to facilitate illicit trade.[37] In her study of the illegal trade in caviar in the EU, Dickinson refers to these as "grey markets" because the trade is a complex blend of legal and illegal sources and actors who are able to exploit legal loopholes. The mixing of legally and illegally sourced caviar has arisen since the tightening of regulations on caviar. In 1998 all species of sturgeon were listed in CITES Appendices; these listings meant that thereafter trade in wild caught caviar became heavily regulated, with a complete ban on trade in eggs from some sturgeon species. In order to meet demand for caviar, a global industry developed to farm it in aquaculture facilities. Over a twenty-year period there has been a significant shift toward farmed caviar, such that more than 90 percent of caviar is now legally produced by aquaculture facilities. But legally produced caviar can be mixed with caviar from illegal sources. For example, illegally sourced caviar from wild sturgeon populations can be

Unlabeled legally produced jars of fresh caviar; which will be repackaged into smaller caviar tins. Photo credit: Hannah Dickinson.

deliberately labeled as farmed from aquaculture facilities and can thus enter legal markets to be sold openly; wild and farmed caviar can be mixed together and sold as legally produced caviar; and since individuals are allowed to have small amounts of wild caviar for personal consumption, these small amounts can also be sold in illegal markets.[38]

One of the challenges in thinking around wildlife crime is defining which activities and species should be included and which should not. Wildlife crime is much broader than just poaching and trafficking. Illegal, unregulated, and underreported (IUU) fishing is dealt with as a separate issue to illegal wildlife trade because it is (primarily) large-scale fishing of high-value commercial species, such as a tuna, salmon, cod, mackerel, shrimp, and abalone. IUU fishing is estimated to generate US$10 billion to US$23 billion annually, and most damage results from the illegal fishing conducted by corporate-owned or state-owned fishing fleets.[39] Trafficking of sturgeon caviar and of glass eels in Europe are considered as examples of the illegal wildlife trade and tend not to be thought of as IUU, even though both are commercially valuable fish.

IUU fishing illustrates the challenges involved in framing and defining wildlife crime, which ultimately shape policy responses. However, if we zoom out and expand the lens through which we understand what constitutes a wildlife crime, it allows us to examine a wider range of framings and practices on the ground.

POACHING AS WILDLIFE CRIME

Poaching is often referred to as a wildlife crime, however the term "poaching" is contested and can be deployed in ways that obscure historical economic, social, and political dynamics. As Lubilo and Hebinck suggest, the term "poaching" is inadequate because it does not capture why local forms of hunting persist, even when such hunting is against the law. They argue that "local hunting" or "local hunters" are more accurate descriptors, as they take account of the ways that rules and laws produce the category of poacher as a criminal.[40]In conservation circles poachers are often presented either as ruthless criminals targeting high-value species and motivated by economic gain or as poverty-stricken individuals who hunt/trap small animals for subsistence purpose. Neither of these images pays attention to how the ongoing legacies of colonialism continue to shape and define who is called a poacher.

One example of poachers being presented as greedy, well-equipped criminals is in the WWF-International report *Big Wins in the War against Wildlife Crime*. It states: "To maximize their profits, poaching networks behave in the least sustainable and most socially, economically and environmentally damaging ways, exacting a huge cost on local societies. The heavily armed gangs attack law enforcement agents, murder civilians, and engage in a wide range of other criminal activities."[41] This framing can obscure the wider context that drives people into the poaching economy in the first place, making it appear as a matter of choice and as motivated by economic gain. This then becomes part of a justificatory framework for swift and even forceful responses, yet this is just one characterization of poachers.

One simple definition of poaching is any hunting or removal of wildlife that is not sanctioned by the state or private owner. Yet it is important to address the wider historical, social, and political reasons for why some activities become defined as illegal while others do not.[42] The history of how hunting became defined as poaching is centrally important to understanding why poaching persists in sub-Saharan Africa. There is a substantial body of research on how historical grievances, poverty, inequality, and the continuing effects of colonization and racism continue to shape and drive poaching.[43] For many communities, the removal of access to land and wildlife via the introduction of wildlife laws during the colonial period remains a source of grievance. These laws fundamentally changed the relationship between people and their surrounding environment by creating protected areas and sport hunting reserves or zoning for settler agriculture, among other things, which resulted in dispossession, eviction, and displacement.

Furthermore, it is important to distinguish between subsistence and commercial poaching while recognizing that these are also imperfect and overlapping categorizations. Subsistence poachers typically target smaller animals for food, rely on low technology techniques (notably the use of traps and snares), and tend to have a less significant impact on wildlife populations.[44] Some forms of illegal hunting for meat straddles both subsistence and commercial-scale activities.[45] Current debates about poaching tend to revolve around concerns about those defined as commercial poachers who are more organized and target economically valuable species (rhinos, elephants, tigers, pangolins) for trade rather than subsistence. These "commercial poachers" may have more advanced technologies, including firearms, GPS units, and mobile phones, and have a more negative impact on wildlife.[46]

Within communities, poachers can be regarded as people who challenge the rich and powerful; they can be accepted, and even lauded, by poorer and more marginalized communities because they breach boundaries regarded as unjust. It is important to understand this contextual politics of the law, as well as the complexities of poaching and of poacher motivations.

Defining poaching is critically important. The switch to describing the illegal wildlife trade as *wildlife crime* renders this contextual politics of poaching invisible. It communicates the idea that perpetrators are criminals, knowingly and willingly engaged in breaking the law for their own benefit, with little regard for the wildlife or the impact on ecosystems. But this partial view of poaching does not address how, in some places, colonialism created poaching as a category of illegal behavior. Furthermore, the use of the term "wildlife crime" also obscures the ways that people may be drawn into a poaching economy: to make ends meet as a result of a lack of other opportunities, as a means of addressing relative inequalities, or because they have been coerced into it by more powerful people and networks in their societies.

LIVELIHOODS AND WILDLIFE CRIME

It is important to address the ways that the (legal and illegal) wildlife trade can be a significant contribution to livelihood strategies in some of the poorest communities in the world. This discussion intersects with the debates on human security. The illegal wildlife trade can shape and have an impact on livelihoods in two ways. First, wildlife losses may deprive those communities of sources of food and income. Second, consuming or trading illegal wildlife products may be an important means of generating income and meeting basic needs. The *EU Action Plan against Wildlife Trafficking* (2016–20) acknowledges this and identifies the need to engage with rural communities to address the root causes of the illegal wildlife trade. Objective 1.2 of the plan is to ensure that rural communities in source countries are more engaged in tackling the illegal wildlife trade. Doing this includes supporting a legal and sustainable wildlife trade in wildlife conservation to ensure that communities benefit more from it.[47]

One good example is the development of baobab trade by PhytoTrade Africa, which invested significant resources in getting the EU to agree to allow the plant to be traded into the EU because the baobab-based products needed to meet stringent consumer safety tests.[48] In a similar initiative, the case of Baobab Products Mozambique (BPM) indicates that communities need a lot of support and assistance to ensure they are able to capture the full value of the products. In the case of BPM, community members were trained to prepare and process baobab products in order to capture more of their economic value.[49] Cooney et al. point out that the structure of the supply chain has a strong bearing on the incentives for conservation and the opportunities for poor people to participate and benefit from the trade.[50] For example, the costs of production may be a barrier to entry. Longer supply chains may mean that benefits are more widely distributed, with fewer returns to communities involved in the early stages of harvest and processing. Concentrated market power may favor or hamper the livelihoods of communities, depending on which stage of the supply chain they are involved in.[51]

Providing economic alternatives for those engaged in unsustainable and/or illegal wildlife trade may be only part of the solution. Property rights governing the use of land and wildlife can have significant implications for the commercial viability of legal wildlife trade, for the incentives for sustainable use, and for the associated livelihood benefits. One case is the legal trade in the yellow anaconda, developed alongside a regional community management scheme in Argentina. The scheme improved wetland management and population sustainability and provided supplementary income to around three hundred local people from La

Estrella Marsh.[52]It is important that the conversation around framing the illegal wildlife trade as a wildlife crime not overlook the complex interrelationships between the trade and sustainable livelihoods. This has become even more important amid calls for tougher regulation and enforcement in the wake of the COVID-19 crisis; researchers such as Dilys Roe have highlighted how increased use of trade bans could be devastating for communities that are reliant on a sustainable and legal trade in wildlife.[53] A focus on breaches of the law/rule breaking, and gaining compliance with the law, excludes these wider issues around the use of plants and animals as part of sustainable livelihood strategies.

CRIMES AGAINST WILDLIFE

Framing illegal wildlife trade as wildlife crime does not adequately encompass the emerging ways of thinking about how crimes can be perpetrated against the environment in general and wildlife in particular. Criminology as a field has begun to grapple with the idea of crimes against the environment as a distinguishable form of crime. Until recently the assumption was that crimes were activities that caused harm to human beings,[54] not that they were harms perpetrated against wildlife or nature more broadly. Green criminology[55] draws attention to the ways criminal activity, and the harms it produces, have negative impacts on nature (see chapter 1).[56] As Ragnhild Sollund suggests, green criminology draws on ideas of ecological citizenship and ecological justice, which underline the duties of humans to coexist with the nature. Further it has explored how legislation accomplishes or falls short of its intended purpose, and what or whom it benefits.[57]

The idea of ecocide also is important here. Eckersley has offered important reflections on the notion of ecocide. In the case of conservation, she raises the question of whether the international community should be concerned about massacres perpetrated against critically endangered species.[58] If the international community has a responsibility toward endangered species, then that raises the possibility of *ecological intervention* and the development of an international environmental court to deal with crimes of ecocide. Humphreys and Smith suggest that invoking notions of force to protect the environment, especially wildlife, is intuitively unacceptable for many.[59] However, they also suggest that animals can and do suffer from state failure, so there can be a moral imperative for international intervention to save wildlife if the state is incapable or unwilling to do so.[60] If the international community does have a responsibility to act to prevent harms, including extinctions, then the notion of a UN-backed force to meet these legal duties to prevent crimes of ecocide becomes a possibility, one supported by a moral case.[61]

Further, prevention of ecocide can be underpinned by the rationale that it enhances security. This is not an entirely abstract notion, and it is part of the thinking in international policy networks. For example, a senior European environmental law enforcement professional commented, "If you look at factories polluting the environment in the vicinity of borders, polluting the ground water . . . if you look at illegal fisheries in the Atlantic Ocean, where resources are being depleted in the long term, there is a security aspect involved because we rely globally on these kinds of resources, if not in this decade then in the coming decades."[62]

There is a clear link drawn between environmental harms and security because of human reliance on nature to sustain itself. This idea ties in with the ways that green criminologists highlight harms against nature.[63] Thinking about wildlife crime in this more expansive manner allows greater critical engagement with the ways poachers and poaching are defined, and it raises the possibility of thinking about how governments and businesses can harm the environment (including as perpetrators of wildlife crime). It is important for understanding how wildlife crime is articulated as a security threat.

SECURITY THREATS: WILDLIFE, ORGANIZED AND DISORGANIZED CRIME

Presenting the illegal wildlife trade as wildlife crime facilitates its framing as a security threat by drawing links between corruption and organized crime networks and other illicit trades in drugs, weapons, and people.[64] As Lavorgna and Sajeva note, a key aspect of the argument that wildlife crime is a security threat is that it is linked to corruption and organized crime.[65] The United Nations Convention against Transnational Organized Crime (UNTOC) defines an organized criminal group as "a structured group of three or more persons, existing for a period of time and acting in concert with the aim of committing one or more serious crimes or offences established in accordance with this Convention, in order to obtain, directly or indirectly, a financial or other material benefit."[66] This definition encompasses a range of different actors and not just the more well-known networks, such as mafias, triads. The UNODC determines that nearly all transnational wildlife trafficking fulfils these criteria.[67]

A wide range of organizations draw the link between organized crime and wildlife trafficking, including the UN Economic and Social Council (ECOSOC), INTERPOL, United Nations Environment Program (UNEP), and the United Nations Security Council.[68] For example, the EUROPOL Serious and Organized Crime Threat Assessment (SOCTA)[69] identifies organized crime groups as engaged in wildlife trafficking.[70] Both UNTOC and UNODC tackle wildlife and timber trafficking, and provide a framework for approaching transnational organized crime.

The UNODC, in particular, addresses the intersections between illegal wildlife trade and organized crime. The role of UNODC[71] is to support member states in addressing environmental crimes, but its initiatives around wildlife crime are still in their infancy.[72] From 2012 all activities to address wildlife and forest crime were brought under the first UNODC Global Program on Combating Wildlife and Forest Crime, managed by the UNODC headquarters in Vienna.[73] As part of that program UNODC produced the first *Global Wildlife Crime Assessment* in 2016, due to its concern that wildlife trafficking was becoming recognized as an area of specialism for organized crime and because the trade itself threatened the survival of several species.[74] The support provided by UNODC under the global program includes legislative assistance, training, and provision of essential equipment. The program aims to benefit police, customs, border officials, forestry and wildlife officials, prosecutors, the judiciary, and community groups.[75]

The International Consortium on Combatting Wildlife Crime (ICCWC) was established in 2010 to address the increasing sophistication and involvement of organized crime networks in wildlife crime, as a well as the fractured and inadequate responses at national, regional, and international levels.[76] It was an initiative of INTERPOL, CITES, the World Bank, the World Customs Organization (WCO), and the UNODC operating as equal partners. Its purpose is to provide coordinated support to national wildlife law enforcement agencies and to such regional networks as the Wildlife Enforcement Networks (WENs). ICCWC developed the Wildlife and Forest Crime Analytic toolkit in 2012 and provided specialized training for national agencies in 2013.[77] In addition, ICCWC supports countries at the national level; for example, it deployed a Wildlife Incident Support Team to Madagascar, Sri Lanka, and UAE.[78] ICCWC also convened the first Global Partnerships Coordination Forum at the CITES CoP17 in 2016.[79]

Drawing the link between wildlife trafficking and corruption is central to the ways that some of the big conservation NGOs articulate the urgency of addressing wildlife crime as a national and global security threat. For example, WWF-International states: "Increasingly involving large-scale, transnational organized crime, the current unprecedented spike in illegal wildlife trade poses a growing threat not only to wildlife but also to security, rule of law, sustainable development, and the wellbeing of local communities."[80] Corruption is certainly an important component that enables wildlife to be poached, smuggled, and sold, but it should be seen as a facilitator of the illegal wildlife trade, not its driver or cause.[81] Corrupt networks may view wildlife as a lucrative resource they can use to enhance their position or gain financial rewards. Corrupt officials or business people can use the profits from wildlife trafficking for patronage or they may allow others access to lucrative wildlife in return for financial returns or to enhance their power and social status. One of the characteristics of these networks is the specialized brokers who provide services (such as provision of false permits), which facilitate trade, including via funneling illegal wildlife products into the legal wildlife trade.[82] Different methods are used, among them money laundering, bribery, use of diplomatic bags for smuggling, and provision of false CITES permits.[83]

The illegal wildlife trade is attractive to organized crime groups because it can be low risk, high reward. The profits can be significant and the chances of getting caught are lower than for other illicit trades.[84] For example, INTERPOL estimated that in 2013 a significant portion of ivory reaching international markets, especially in Asia, was derived from elephant populations in Tanzania and that the increase in large-scale shipments indicated the participation of organized crime, with trafficking syndicates operating in multiple countries simultaneously. These crime syndicates sourced ivory from several hundred elephants for each shipment, and they have been primarily responsible for the drastic decline of African elephant populations since the late 2000s.[85] However, as Titeca argues, there is little actual understanding of exactly how ivory is sourced and transported through trading networks in Africa, which then join up with organized crime networks.[86] Hence, there is a question mark over whether the illegal wildlife trade is prosecuted by organized crime networks, or merely by networks that include criminal elements, or even by forms of disorganized crime. As Massé et al. suggest, the portrayals of organized criminal groups and people involved in illegal wildlife trading are often simplistic, and they further entrench the idea that most or all wildlife crime is organized and violent. While there are cases where illegal wildlife trade does demonstrably involve organized crime groups, the evidence for certain trades assumed to be driven by organized crime is anecdotal, lacking entirely, or it conflates organized crime groups with crime that is organized by different networks of traffickers that do not meet the internationally accepted definition of organized crime.[87]

Much attention has been focused on the role of organized crime, but the illegal wildlife trade is also carried out by disorganized, eclectic, and temporary networks.[88] Felbab-Brown suggests that there are three distinct interdependent groups involved in sustaining the illegal wildlife trade: consumers, middlemen, and suppliers.[89] However, the networks involved in the illegal wildlife trade are not necessarily as neat as this, and different types of actors can move in and out of the legal and illegal trades. The challenge of coming to grips with whether wildlife trafficking is conducted by organized crime networks, or not, is central to USFWS thinking about the illegal wildlife trade. Some members of the service actually view wildlife trafficking as quite a disparate, disorganized phenomenon, commenting, "It depends on what your definition of organized crime is, if you use the classic sense you can see it in Africa and South America. A lot of times it is disorganized crime. The higher the product goes in the chain to its destination the more organized the criminals are."[90]

The role of disorganized crime is an important feature, frequently overlooked in more mainstream and high-profile accounts. As another official stated, "In Asia, you have organized crime more in the big-ticket items, not just ivory and rhino horn. But at a more local level, the high-volume items, they are not worth a lot but there are a lot of them, like birds in Indonesia or civets. Smaller criminal networks bring them out of the forest into the market and then they are the single source and they give them out to traders."[91]

It somewhat complicates the story when illegal wildlife trade is a security threat carried out by organized crime networks, which undermine governments and the rule of law. The focus on transnational organized crime is a more readily intelligible and acceptable way of explaining the persistence, scale, and value of the illegal wildlife trade at international forums; placing the blame on shadowy organized crime networks is much less politically sensitive than identifying the roles of corrupt elites, corporations, or an ill-defined, diffuse, and disorganized network.

Echoing arguments by Wyatt about the roles of disorganized, fluid networks engaged in the illegal wildlife trade, Titeca's study of ivory traders in Uganda contests the notion that wildlife trafficking constitutes organized crime.[92] Some ivory traders in Arua moved ivory across the border from DRC to Uganda in small amounts on the backs of bicycles, storing it until they had a large stockpile. Elements of the military also contracted civilian truck drivers to transport ivory across the border as part of other consignments of goods. The traders interviewed by Titeca in Arua and Kampala mentioned several different nationalities as important buyers who traded ivory into international networks: Pakistani, South African, Chinese, and Korean. The range of organizations that ivory traders are in contact with is also instructive. Titeca gives the example of Adam, based in Kampala, who had been involved in ivory trading for more than fifteen years and was in regular contact with corrupt officials from the Uganda Wildlife Authority, the Tanzania Wildlife Authority, elements of the Ugandan army operating in Central African Republic, and Burundian traders, as well as small-scale traders from the border town of Arua.[93] Adam's close links with a particular Ugandan military commander, who had long experience of working in DRC, were extremely important to his ivory-trading business.[94] The picture presented here does not map well onto the idea of a highly organized network of ivory traders, linked in to global organized crime networks that undermine governments and the rule of law.[95] Generalizations that frame all illegal wildlife trade as serious organized crime do not capture the messier realities on the ground, including how networks are structured and the often informal, mundane, and opportunistic ways in which a lot of wildlife crime occurs.[96] This creates blind spots and omissions in policies to tackle the trade.

INTERSECTIONS BETWEEN WILDLIFE TRAFFICKING AND OTHER ILLICIT TRADES

One of the most common arguments made about the illegal wildlife trade as a security threat is that it is integrated with other illegal trades, including drugs, antiquities, weapons, and people. This configuration is sometimes referred to as "crime convergence." As a result there are calls to learn the lessons of the international narcotics trade, especially the limitations of policies and the absence of silver bullets.[97] The European Union Agency for Criminal Justice Cooperation (EUROJUST) report on environmental crime points out that in cases of wildlife crime the most common links are to offenses related to corruption, the fraudulent obtaining of licenses or forgery of the latter (including customs official documents), money laundering, and drug trafficking.[98] The Organization for Economic Cooperation and Development (OECD) Taskforce on Countering Illicit Trade aims to coordinate international expertise in mapping illicit markets to understand trafficking in order to inform more effective policies to tackle it. Its main foci are drugs, arms, persons, toxic waste, counterfeit consumer goods, and wildlife. It addresses the wildlife trade as part of wider concerns about trafficking, and it has begun to aggregate publicly available data in order to map the main trafficking hotspots.[99]

In the US, the argument was made that illegal wildlife traffickers were utilizing much longer-standing drug trafficking routes and networks. In recognition of this, in 2015 wildlife crime was added to the definition of transnational organized criminal networks in the US National Defense Authorization Act.[100] As one US government official noted, "There is a clear and present danger to governments and their ability to govern, posed by very lucrative and very powerful and very connected industries, which show a great deal of convergence between wildlife trade and other forms of illegal trafficking."[101] Also in 2015, the US House of Representatives passed the Global Anti-Poaching Act, giving wildlife crime the same criminal status as drug and gun smuggling, which allowed for increased penalties. It also made wildlife trafficking a liable offense for money laundering and racketeering. Further, in the 2016 Intelligence Authorization Act, the US House of Representatives mandated the director of National Intelligence to report on wildlife trafficking networks and targets for disruption.[102] Each of these actions elevated the illegal wildlife trade to a much greater level of importance, and with that came more attention, funding, and policy commitments from the US government.

The EU approach is to link policies on the illegal wildlife trade to other illegal trades, believing that tackling the illegal trade in drugs, people, and stolen cars can help result in prevention of the illegal wildlife trade. For example, in 2015 the European

Parliament and European Council also passed the fourth Anti-Money Laundering Directive (EU) 2015/849 on the prevention of the use of the EU financial system for the purposes of money laundering or terrorist financing.[103] The action plan in the Directive aimed to develop a list of high-risk third countries in 2016, countries that might overlap with important sources and destinations of illegal wildlife products. For example, the EU has been providing support to the Kenyan government to implement best practice to prevent money laundering and financing of terrorism; Kenya is also an important source of illegally obtained wildlife products (notably ivory).

The view that wildlife trafficking mirrors other illicit trades, such as drugs, was clearly stated by one MEP: "A lot of these are the same gangs, same groups, same processes, money getting moved around in the same ways. In fact, it got a lot easier to do wildlife trafficking, they moved into human trafficking because drugs were getting too difficult and dangerous and court sanctions were getting too big."[104] However, a senior environmental law enforcement official was keen to point out that while there was an assumption that wildlife trafficking was linked to other forms of organized crime, the evidence was not conclusive. The official continued by saying: "The criminals are so specialized they trade only with particular species, they don't traffic in animals and plants in general. If you bear that in mind then a good question is to what extent are they really involved in other forms of crime? There have been cases where this has been reported but it is scientifically difficult to say if you have one or two cases that this is a common phenomenon."[105]

Drawing parallels with the drugs trade, and especially with the US-led War on Drugs is tempting, but although they do share some structural similarities, the narcotics trade and the illegal wildlife trade are not analogous. The key differences between them will inevitably shape effective policy responses, which draw on the experiences of the drugs trade. In particular, the illegal wildlife trade is shaped and sustained by local institutional and cultural contexts, which determine who benefits and who is affected negatively by it. This aspect needs to be addressed as part of any policy response to the illegal wildlife trade. One good example is that some scientists recently called for conservation biologists to apply the lessons learned from archaeology in describing and publicizing their new discoveries. In archaeology, it is common practice not to disclose the precise geographical location of new discoveries when they are published and publicized. This is to prevent theft and to ensure new discoveries do not fuel the illicit trade in antiquities.[106]

While it is important to address the role of criminal networks, the ways that states themselves can be engaged in trafficking is overlooked. It goes well beyond the corrupt activities of individual officials and instead might be thought of as systematized trade using the powers and resources at the disposal of states to capture profits from the illegal wildlife trade. A good example comes from the illegal trade in valuable hardwoods. The economic value of illegally harvested and traded rosewood has shaped the dynamics of the state in both Vietnam and Madagascar. The rosewood trade from Laos to Vietnam is an important means of accumulation for a complex and fluid range of actors. Such actors easily move between state and private sector because they have overlapping social spheres, which involve expectations of reciprocity across social, political, and economic domains.[107] Similarly, the development of a group of rosewood barons after the coup d'état in 2009 and before the return to elections in 2013 in Madagascar, have shaped the governance structures of that state. Corrupt patronage networks were able to maintain a significant degree of political power. A study on the political ramifications of the illegal rosewood logging published in 2018 in *Political Geography* was written anonymously, which reflects the significant degree of risk to the author and research participants. What the research demonstrates is that the rosewood barons have been able to develop high levels of support among their constituents, which enabled them to be voted into office in the 2013 elections. They connected small-scale operators working in logging camps in the forests of northeastern Madagascar with Chinese ships waiting offshore to transport the logs directly to Asia. This network produced what the author calls a rosewood democracy, which continues to shape the nature of governance structures in Madagascar today.[108]

A fuller understanding of how the illegal wildlife trade shapes politics in Madagascar challenges the idea that the rosewood trade is driven by corruption carried out by organized crime networks and that therefore the solution lies in better law enforcement and good governance.[109] This example illustrates why it is important to develop a more nuanced understanding of the wider context of the illegal wildlife trade, and why defining as a security threat, matters.

Defining the illegal wildlife trade as a wildlife crime, a serious crime, and a security threat fundamentally reshapes policy toward it. Thinking about the illegal wildlife trade as a matter of crime and security prompts responses that are focused around better law enforcement, enhanced compliance, and mitigation of security risks. This shift moves the focus of conservation policy away from seeing the illegal wildlife trade as produced by inequalities between wealthier consumers and poorer suppliers, or as a wider structural issue of development, lack of opportunity, and the dynamics of the global economy.

It raises important questions about how we define a wildlife crime. The ways that poaching and trafficking are framed informs understandings of who or what can be perpetrators of wildlife crime. The idea that the illegal wildlife trade is carried out by criminals, organized crime networks, and shadowy groups operating outside the law has facilitated the growing integration of conservation and security. This framing can obscure some of the underlying drivers of the illegal wildlife trade, such as demand from the world's wealthiest communities. Further, it is important to address why some people continue to engage in illegal hunting; it may be because they have no alternative opportunities or because they do not recognize the legitimacy of colonial laws that forbid access to or use of wildlife in the area. Using a political ecology perspective can help to reveal these omissions and blind spots. Framing the illegal wildlife trade as serious, organized crime and as a security threat changes approaches to conservation policy.

3 WAR FOR BIODIVERSITY

Conservation is sometimes cast as a "war for biodiversity." In this chapter I examine how that can translate into different kinds of practices on the ground. The sense of urgency, coupled with the framing of illegal wildlife trade as a serious crime, has ushered in a fresh phase of the war for biodiversity. There has been greater support for militarization of conservation, especially in sub-Saharan Africa. For some in the conservation sector, the need for more forceful intervention to protect national parks and species is a necessary response in the face of intense pressure, most especially in areas of armed conflict and/or where state capacity is limited. Proponents argue that bad examples of militarization (torture, killings, alienation of local communities, and so on) are the result of poor practice or that they are isolated, unrepresentative cases perpetrated by the "rotten apples" in the barrel.

The shift toward more militarized practices in tackling the illegal wildlife trade has raised some central questions. Is the historical context of engagements between militaries and conservation important for understanding current approaches? Is militarization displacing or combining with approaches rooted in Community Based Natural Resource Management (CBNRM) or neoliberalism? Why has the war for biodiversity re-emerged and why are there concerns about human rights abuses?

BATTLING FOR BIODIVERSITY

Referring to the current steps tackling the illegal wildlife trade as a "war for biodiversity" is a very slippery concept. It is even more difficult to define its practices. First, there are the discursive practices of conservationists, media, and contributors to social media, who refer to the need to save species as a "war," which indicates that it is a battle (militarized or not).[1] Second, there are the related physical practices of war to save wildlife, which range from surveillance to shoot-to-kill orders for suspected poachers.

The arguments in favor of militarized conservation are very strong. In brief, militarization is presented as the most effective response to an emergency situation; that conservationists have no other option to defend wildlife; and that more militarized anti-poaching results in recovery of wildlife populations.[2] Indeed, even though officially WWF funding cannot be used to purchase weapons, Luis Arranz, WWF manager at Dzangha-Sangha National Park in Central African Republic, openly states, "I need guns to protect the people that protect the elephants. If they meet poachers and are unarmed, they will die."[3] Hilborn et al. have shown that heavier levels of enforcement of protected areas can result in positive outcomes in terms of species conservation.[4] Further, the precise practices of militarization are tailored according to location, species, and available resources.

Militarization takes many different forms, and it is not a singular thing. It can involve the authorization for conservation authorities (usually rangers) to use deadly force to protect themselves and wildlife; it can also include the development, adoption, or adaptation of military-style thinking, tactics, and training and/or the provision of military equipment as a means of achieving conservation objectives.[5] More broadly the militarization of conservation parallels the wider ways in which policing has become more militarized, blurring the boundaries between law enforcement, military, and security.[6] Wilson and Boratto point out that in turning toward security and heavier forms of law enforcement, conservationists have not paid sufficient attention to the evidence from policing that indicates it has little effect on reducing crime rates and on reducing reoffenses; further, they contend that such approaches can be harmful to people, communities, and states.[7]

There is a clear politics of race in the war for biodiversity, which is largely rendered invisible in the positive coverage of the advantages of militarized approaches. Marijnen and Verweijen call this "white authorized green violence."[8] This refers to the use of force, which is defined as legitimate when whites/westerners control and order it (but do not necessarily execute it). For example, conservation in Virunga National Park is characterized by the production of consumable images of warscapes: heroic (white) military trainers and (black) armed park guards figuring as engaged in an important crusade to save wildlife. This is illustrated by the popularity of the Netflix movie, *Virunga,* which is accompanied by the tagline "Conservation is war" and images of a heroic (armed) ranger standing alone in the landscape, guarding a single mountain gorilla at his feet. The film was supported by the philanthropic Di Caprio Foundation (founded by the actor Leonardo Di Caprio) and is described as the "true story of rangers risking their lives to save Africa's most precious park and its endangered gorillas."[9] There is no doubt that those involved in conservation in Virunga face intense pressures, and that they are deliberately targeted by armed groups in the area as well as caught in the crossfire when defending civilians from armed groups (as in the case of thirteen rangers killed in April 2020).[10] However, in the context of complex conflicts, it is too simplistic to separate out conservation practices as good versus other armed groups, 'the bad guys.'[11] The ways that conservation problems are presented as urgent, as emergencies, with their symbols of good and bad guys have a material impact on the ground.

The war for biodiversity, like other wars, has its human as well as animal casualties. The figures on the losses of elephants, rhinos, tigers, and pangolins are all readily available. What is less clear is the human toll of the renewed war for biodiversity. It is very difficult to obtain definitive numbers of people killed and injured in ongoing operations. However, there are some telling examples that illuminate the scale of the problem. In 2014 several

Display at the IUCN World Parks Congress, Sydney, Australia, 2014. Photo credit: Rosaleen Duffy.

conservation NGOs, international organizations, and news media began reporting that one thousand rangers had been killed in anti-poaching operations in a ten-year period. This was a central argument for organizations and individuals pushing for a more forceful response to rises in poaching. The figure of one thousand originated from the Thin Green Line Foundation,[12] and it was taken up by several organizations, including United for Wildlife,[13] IUCN,[14] and WWF-International,[15] among many others, to draw attention to the pressures faced by rangers in the field. The figure is likely to be an underestimate because it is based on the number of families the foundation assists. Further, the foundation has no data for a number of states, including Nigeria and Somalia.[16]

It is not just rangers who have lost their lives in the escalating conflict (or war) over wildlife.[17] Those identified as poachers have also been killed in anti-poaching operations, and countless have died while on poaching expeditions as a result of exposure, injury, and encounters with wildlife. It is extremely difficult to obtain figures on other deaths associated with the war for biodiversity—figures are simply not collected in a systematic way, or not collected at all. In one of the few studies that tackles this question, Lunstrum suggests that more than three hundred suspected poachers were killed between 2008 and 2013 in Kruger National Park alone.[18] In 2018 there were clear tensions between Namibia, Botswana, and Zimbabwe over the approach to poaching by the Botswana government under President Ian Khama (2008–18). When he was elected in May 2018, the new president of Botswana, Mokgweetsi Masisi, suspended shoot-to-kill and removed military grade weapons from anti-poaching units operating near the country's borders. The Namibian press reported that thirty Namibians and twenty-two Zimbabweans had been killed as suspected poachers in Botswana.[19] It was not the first time that aggressive anti-poaching and shoot-to-kill sparked tensions between neighbouring states. There were similar tensions between Zimbabwe and Zambia in the 1990s because of Zimbabwe's militarized approach to rhino and elephant protection in the Zambezi Valley.[20] Such escalations in the use of force, and the injuries and deaths associated, with them require more exploration and explanation.

In order to do this, it is important to sketch out the historical context of the war for biodiversity, including historical connections between the military and conservation, the faith in fortress conservation, and the development of alternatives like CBNRM and market-based conservation. The current integration of security and biodiversity conservation builds on these important historical factors and has developed into a new phase of conservation, characterized by a much fuller integration of global security concerns and the deployment of approaches more commonly associated with military and counterinsurgency strategies. Büscher and Fletcher refer to this new phase as "green wars" because the current scope, scale, and vocal justification of a violent defense of biodiversity is unprecedented in conservation.[21] But militarization is just one expression of the integration of conservation and security, which is a much broader set of developments. It is, however, the most high-profile aspect of the ways conservation and security are becoming more integrated, and it tends to generate the most intense debates and disagreements over the ethics and efficacy of the approach.

THE 1980S AND THE RISE AND RISE OF THE WAR MODEL

Current calls for more forceful responses to poaching are often presented in an ahistorical way, as if they are new. Yet current strategies build on a long history of engagement between militaries and conservation, including military-style enforcement. For example, the US Cavalry managed Yellowstone National Park as its first park rangers, from 1872 to 1916, and the national parks were used for military training during World War Two. National parks have also been used by militias, such as Fuerzas Armadas Revolucionarias de Colombia (FARC) in Colombia, to train and launch operations. As such these parks have seen significant levels of intervention by state security agencies. Some parks, for example, Big Bend National Park, which follows the US-Mexico border, have even been identified as security buffer zones between states. This use builds on historical models of fortress conservation that relied on separating people and wildlife to create a network of national parks and hunting reserves for conservation, tourism, and sport hunting. This fundamentally changed relationships between wildlife and human communities, creating islands of people-free wilderness that needed to be preserved and actively defended.[22]

Thirty years ago, the world of conservation was gripped by the wildlife wars to save elephants and rhinos. In the 1980s, the demand for ivory saw elephant numbers plummet by 50 percent, from approximately 1 million to 500,000, in a decade. Several countries developed new, more deadly, anti-poaching programs to secure elephant populations: Operation Uhai in Tanzania and Operation Stronghold in Zimbabwe. Similarly, the declaration of shoot-on-sight policies in Malawi, Central African Republic, and famously in Kenya by President Daniel Arap Moi and Richard Leakey, director of Kenya Wildlife Services (KWS), in 1988.[23] Richard Leakey's term as head of KWS was characterized by growing militarization of protected areas and use of deadly force against poachers.[24] The KWS promoted this approach as necessary to prevent elephant extinction. A de-escalation occurred only with the sharp drop in poaching and trafficking following the ban on ivory trading in 1989.

Back in 1989 it was the intensity of concern, shared by governments, NGOs, international organizations, and by the general public (especially in the US and Europe) that built global pressure for an international ban on the ivory trade. In 1989 CITES held its Conference of the Parties in Lausanne, Switzerland. The meeting attracted interest from conservation NGOs and from the wider public pushing for a ban on the ivory trade in order to solve the elephant poaching crisis. At the meeting, members voted to move African and Asian elephants to the Appendix 1 list, which translated into a total ban on international trade in elephant ivory. There were some exceptions, which arose as a result of disagreements between the main elephant range states in Africa. Several Southern African states argued (and still do) that they had too many elephants, not too few, and that they were being punished for failures in elephant conservation in other regions. As a result, Southern Africa's elephants remained on Appendix 2, which permitted a strictly controlled and monitored internal trade in ivory plus sales off stockpiles.[25] The ivory ban was credited with reducing poaching and trafficking of elephant ivory in the 1990s, allowing elephant numbers to recover until poaching rates began to rise again after 2008. Apart from the ban on ivory trading, conservation in the 1990s was characterized by the expansion of alternative approaches to conservation that were more focused around community-led initiatives and sustainable use.

CBNRM TO CBCM: MILITARIZATION AND COMMUNITIES AS ENFORCERS

CBNRM is often regarded as the opposite of militarization of conservation. CBNRM rose as a counterpoint to fortress conservation, then faced criticism and challenges, and culminated in a shift back toward more forceful methods of conserving wildlife. But CBNRM has not been completely displaced by the militarization of conservation. Instead, it has been blended with militarized approaches, as a result of the sense of urgency around more recent wildlife losses. I call this community-based militarized conservation (CBMC) in order to capture the ways that CBNRM and militarization have been integrated, even fused together in interesting and sometimes counterintuitive ways.

In the 1990s a new wave in conservation favoured more community-oriented approaches. CBNRM was (in part) prompted by criticisms of the fortress model of conservation and the use of coercive approaches to defend and maintain protected areas. Its proponents argued that local communities should have stewardship of biodiversity and natural resources through processes of participation, empowerment, and decentralization.[26] CBNRM was designed to be a more socially just and locally acceptable means of developing successful conservation models. There were several high-profile success stories of CBNRM in the 1990s, including Communal Areas Management Plan for Indigenous Resources (CAMPFIRE) in Zimbabwe and the Administrative Management Design Programme (ADMADE) in Zambia. Such schemes were not confined to sub-Saharan Africa; CBNRM projects could also be found in Nicaragua, Philippines, Nepal, Madagascar, and Costa Rica.[27]

CBNRM was central in conservation in the 1990s and was backed by global conservation NGOs, the World Bank, and by government wildlife agencies as means of delivering conservation while alleviating poverty and addressing development. It is an incentive-based scheme which relies on creating local support for conservation. Proponents of CBNRM argue that it offers a win-win scenario: it then contributes to conservation and to development, and local communities gain economic benefits from wildlife. CBNRM schemes are intended to generate benefits from living with wildlife, usually in the form of distribution of meat from wildlife culls, distribution of revenues for utilization of wildlife (via tourism and trophy hunting), and employment in conservation activities and related wildlife tourism sectors.

CBNRM faced growing criticisms in the 1990s that it was in fact top-down and a means by which the state extended its networks of surveillance and informal policing.[28] For example,

Neumann details how rural communities in Tanzania were drawn in to community conservation initiatives and became the eyes and ears of the state. Local groups reported on illegal hunting and resource use, but this conservation technique was practiced in ways that allowed greater levels of state control over lives and lands.[29] Dzingirai suggested that the development of these new community-oriented schemes constituted a kind of new scramble for African lands under the guise of progressive conservation.[30] In 2017 Mbaria and Ogada called wildlife conservation in Kenya the "big (white) conservation lie." For them, claims to community engagement simply facilitate and maintain highly unequal and unjust land distribution, as well as wider forms of social, political, and economic inequality in Kenya.[31] Critics contend that the move to create large-scale conservancies acts to disenfranchize, restructure, and discipline communities in order to secure lands for wildlife conservation. Meanwhile the conservancy model fails to address the aspirations and interests of the communities living within them.[32]

While the CBNRM model faced increasing criticisms, there was also a resurgence in arguments in favor of a return to protectionist and coercive forms of conservation in the mid-2000s.[33] The rationale was that CBNRM was failing to stop declines in wildlife, and so a return to more robust forms of protection was required. This thinking gained greater traction as the rates of poaching of charismatic species, like elephants and rhinos, began to rise again from approximately 2008. CBRNM was persistently criticized as incapable of solving poaching in the long term, especially commercial-scale poaching for lucrative wildlife products like ivory, tiger bone, and rhino horn, because communities could not tackle organized crime networks and well-equipped poaching gangs. This argument has been especially strong where conservation authorities have to operate in the context of wider armed conflict.

In the face of rising poaching, CBNRM has been adopted and adapted in ways that make it compatible with more militarized forms of conservation enforcement. Conservationists articulate the need for greater levels of enforcement as a means of protecting the wildlife as assets (financial or otherwise) for the community.

Furthermore, there is a growing number of examples of conservation initiatives in which community members are provided with training, equipment, and employment in anti-poaching patrols. This builds on a much longer-standing narrative of communities as custodians, stewards, and protectors of wildlife. It also intersects with the argument that employment in conservation can turn poachers in to game guards, by bringing former poachers into the formal economy via waged employment in the conservation sector. In short, their poaching skills and knowledge can be harnessed in the service of wildlife protection. One UK-based conservationist who works with partners in Kenya was keen to point to the benefits of the Northern Rangelands Trust model, stating that they "always work with the Kenya Wildlife Services, Police, the General Support Unit, or the Anti-poaching Unit. . . . I think that's a good model, militarization with care, preferably done by local people and under the authority of the government."[34] Further, a former member of the military who trains rangers in South Africa pointed out that it was essential to work with local communities because they are critical to the success of any anti-poaching initiative.[35] In contrast, a consultant to the US government argued that the interests of local communities actually tended to be overlooked when conservationists develop militarized approaches: "I am interested in understanding what the relation is between land and resource rights on the part of communities, and wildlife conservation. That is including prevention or resistance to poaching. That is part of the security equation that tends to get overlooked . . . the security of the communities and their well-being."[36] Within conservation circles there is little consensus on the value of introducing more militarized techniques into community-based initiatives.

The integration of CBNRM and security-oriented approaches are expressed in new forms of community policing in conservation as well. For example, in Kenya the Tsavo Trust mobilized the arguments around CBNRM and community-oriented policing as a rationale for their Stabilization through Conservation, or StabilCon, program, launched in December 2014. Tsavo Trust aims to recruit and train anti-poaching units from within the local community to enhance the physical security of wildlife and communities in at-risk areas. They use intelligence gathering and development of information networks to develop the program.[37] The program had full support from the governor of the Tana River County and from the Kenyan government. Ian Saunders, chief operations officer for Tsavo Trust, has a background in the UK military, having served in Northern Ireland and Afghanistan. His aim is to apply those lessons to conservation and to extend those skills into communities to develop militarized forms of wildlife conservation. He was also appointed as the security adviser to the governor of Tana River County.[38] We are now seeing this kind of shift in conservation, in which the wider lessons or approaches from militaries and from policing are becoming more common in the face of growing concerns about how poaching and trafficking may lead to wildlife losses and/or contribute to destabilization of regions or governments.

This is one example. There are many others developing in places where there are concerns about rates of poaching. Existing community-led schemes have had to adapt as the situation changes on the ground.[39] One conservationist working on what might be thought of as a militarized form of community-based conservation commented, "Sometimes it is necessary to use military approaches, but it must be done sensitively and appropriately. The communities must be on board and ideally give their approval. . . . [Our] project is based on a community-based natural resource management systems anyway, and the anti-poaching has come a lot later as a sort of last resort."[40]

One community-oriented conservation project, which operates in a zone of intense armed conflict in West Africa, offers interesting insights into how CBNRM can evolve in areas of insecurity. The project began as a means of supporting local communities in managing water resources, elephants, and grazing. But as the government withdrew from the area and in the face of armed insurgencies, the project had to adapt to a new, much more insecure and violent context. Initially the community engaged in intelligence gathering, surveillance, and anti-poaching patrols. As the levels of conflict increased between forces of the government, the UN peacekeepers, and different nonstate armed groups, it became too risky for local community members to be so actively involved. A member of the project stated: "At that stage (2015) the [armed group] hadn't come back so it wasn't dangerous for the local young men to give intelligence or information on poaching. That has happened over the past year. The [armed group] have come back and we now totally divide—so that the local community is totally separate from what the anti-poaching unit does. . . . We had the community aspect but it wasn't enough to stop the poachers. So, the idea was to start a government anti-poaching unit, but there was no government conservation capacity."[41]

In this case there had been a slow shift as a result of the changing security situation on the ground. The project moved from CBNRM, to intelligence gathering and information sharing by community members, toward completely separating from the community the anti-poaching function, which was handed over to the national army instead. This means that some aspects of conservation practice are implemented by a state agency with an (at best) tense relationship with people living in the area.

There is an argument that such operations can provide wider security for wildlife and people in sites of armed insurgencies. Both conservationists and security agencies have argued strongly that conservation is able to provide security services for local communities in areas of conflict, political tensions, or lack of state capacity. One US official, remarked to me, "We are seeing in a number of cases these protected areas in the CAR context that are providing where the state is absent, they are the only semblance of law enforcement around. So, they are providing security for the wildlife in the protected area but also providing security for the communities adjacent to or in the protected area."[42]

Communities are not necessarily opposed to protected areas or enhanced law enforcement and anti-poaching operations. In fact, communities can regard them positively. This intersects much more closely with the human security approach (see chapter 1), in the sense that protecting wildlife can have knock-on effects of creating more security for people living near or with wildlife. As one conservationist running a community-oriented, but militarized, project in a conflict zone stated, "The communities themselves were requesting armed support, that is important to say. They wanted it for anti-banditry." A co-worker added, "And we have these pictures of [religious leader] blessing the anti-poaching unit leaders, having been given goats and things because they were pleased to see them. It is how you do it. As with all conservation tools it is how it is used."[43]

As Hartter and Goldman have shown in their research on two agricultural communities living near to Kibale National Park in Uganda, this is not an isolated example.[44] Local communities can request and support strong enforcement and even militarized conservation as a means of increasing security in their area. For example, a study by Kelly and Gupta showed how protected areas that were once considered bad and inimical to local community interests could be construed as good at different and specific historical moments.[45] People living near Waza National Park in Cameroon hoped that greater levels of law enforcement by park agencies would provide greater security and protection from militias in the area.[46] The highly controlled model of fortress conservation used to manage the protected area was replaced with gradual withdrawal by state authorities in the latter half of the twentieth century. By the early 2000s, local communities noted the rising rates of violence and theft as armed groups moved into the park and used it as a base for kidnapping, cattle theft, and poaching. Over time local communities recognized and acknowledged the importance of park guards in preserving their physical, food, and material security and reminisced almost fondly about the times strong-handed park managers were in place.[47] Furthermore, Lombard and Tubiana also point out that once conservation authorities have developed militarized and security-oriented approaches to protecting wildlife or managing national parks, it can be very difficult to change course; once rangers are armed, the fear of removing arms or relaxing the security can in effect "trap" conservation in militarized approaches because the consequences of any relaxation could be more negative for wildlife and people.[48]

Therefore, while there are substantial criticisms of the militarization of conservation as inimical to community-based conservation or community interests, it is important not to assume there is always a clash of interests when conservation takes a more militarized or security-oriented turn. There is a rich and complex patchwork of outcomes from militarizing conservation, including the ways that it can be folded into CBNRM approaches to create new forms of CBMC. In addition, the tensions between fortress conservation, militarization, and community-led approaches are not the only elements we need to understand in order to explain the development of the current security-oriented phase of conservation. We also need to reflect on the ways that the neoliberal conservation phase has shaped and facilitated the security turn.

NEOLIBERAL CONSERVATION

The integration of security with biodiversity can also be partly explained as the result of the tandem processes of the neoliberalization of nature and the development of a wider global context characterized by fears about security. The current security-oriented phase builds on the market-based approaches to conservation that developed from the late 1990s. The interest in security-related approaches has not swept away CBNRM and neoliberal conservation, but instead these have been folded into a security phase in interesting and innovative ways. The need for enhanced, and even militarized, conservation enforcement is articulated as necessary to protect wildlife as an important economic asset for states and local communities alike.[49] This argument is part of the current debates about the impact of the COVID-19 pandemic in Africa, that the loss of tourism revenues due to the collapse in visitor numbers means that wildlife will face increased pressures from poaching and encroachment on areas set aside for wildlife.[50] This thinking has been facilitated by the development of more market-oriented approaches in conservation over the past twenty years. It is important for understanding why NGOs embraced the security phase, and why private sector security companies have become much more engaged in conservation initiatives.

In the late 1990s, conservation shifted toward a greater engagement with market logics, private sector actors, and corporate sponsorship. These developments sparked a debate about whether nature itself was becoming neoliberalized and whether conservation was indeed being refashioned by it.[51] The debate about neoliberal conservation is well-covered in academic studies so here I will just sketch out its key features.

The neoliberal phase enhanced the levels of integration between conservation and wider market dynamics. Nature was commodified, valued in monetary terms, and made amenable to trading.[52] This trend is discernible in approaches to conservation that rely on payments for ecosystem services (PES), the Economics of Ecosystems and Biodiversity (TEEB), the development of carbon markets, and the promotion of green economy. Similarly, tourism/ecotourism, global markets, and trophy hunting were promoted as ways to make wildlife "pay its way."[53] Conservation NGOs engaged in a process of bringing biodiversity into the realm of economic valuation. For example, the Natural Capital Project brought together the Nature Conservancy (TNC), the World Wide Fund for Nature (WWF), and Stanford University to develop systems for the economic valuation of nature. These shifts were underpinned by a philosophical approach to conservation, which Kathy McAfee refers to as "selling nature to save it."[54]

At the same time, the private sector and philanthropists were becoming more involved in conservation. For example, Conservation International established strategic partnerships with some of the world's biggest companies: Starbucks, Disney, DreamWorks, McDonald's, Coca-Cola, and Walmart.[55] Philanthropists saw conservation as an attractive option, setting up new, private protected areas and ecotourism initiatives, and providing major donations for wildlife conservation.[56] However, the financial crisis (2007 onward) sparked a concern that funding for conservation from the private sector, international organizations, and governments would be reduced. This possibility drove conservation NGOs to seek new sources of support, and because of growing global security concerns it seemed more support would come for activities that could be branded under this banner. These two dynamics produced an environment in which the claim that conservation could contribute to global security was potentially very (economically) valuable indeed.

The turn toward security in conservation displays some continuities with the neoliberal phase, but it also represents some significant breaks with it. For example, the shift has been accompanied by the campaigns against legalization of rhino horn and ivory trades, as well as by claims that wildlife must be conserved as an asset valuable for the tourism industry. Biggs et al. have usefully summarized the two main positions on the ivory trade, an issue they regard as deadlocked in a bitter split over the best means of conserving elephants.[57] On the one hand there are governments and NGOs, especially in Southern Africa, that call for a legalized ivory trade to conserve elephants. Broadly their argument is that since there is a market for ivory, and well-managed (even increasing) elephant populations in some countries, a regulated and sustainable trade could generate much-needed income for their conservation. This can be seen as a market-based solution for elephant conservation.

However, this position faces fierce opposition. The opposing argument is that the only way to secure elephant populations against poaching is to maintain a strict ivory trade ban. Those opposed to legalizing the ivory trade argue that it will be impossible to control the trade, that it would allow unsustainably sourced illegal ivory to be laundered through the legal trade, and that any such trade will have devastating consequences for elephant populations worldwide.[58] Currently, the international trade in ivory is banned under CITES (with some very specific exceptions).

While CITES is the regulatory instrument banning all international trade, several countries (USA, China, UK, and France) have instituted their own domestic ivory trade bans to strengthen legal instruments covering the trade in 2016–18. Therefore, it would not be accurate to argue that neoliberal conservation has been (and continues to be) all-pervasive. While the security phase facilitates some forms of market-based approaches to conservation, it constitutes an obstacle to others. The shift from the neoliberal phase to the security phase has produced both continuities and contradictions.

Conservation exists in a competitive arena and needs to attract funding. Being able to make the argument that conservation contributes to security allows conservationists to tap into the greater levels of funding available in that sector. But that is only part of the explanation. It has suited the interests of militaries, national governments, international organizations, and private military companies to argue that biodiversity losses, and especially wildlife trafficking, constitute global security threats.

THE REBOOT OF THE WAR FOR BIODIVERSITY

By the late 2000s, the rises in poaching, especially of iconic and charismatic species like elephants and rhinos, produced a mood shift. Conservation NGOs, range states, donors, international organizations, and private sector operators drew attention to the need for urgent action to tackle poaching and the illegal wildlife trade.

The US government has taken the illegal wildlife trade much more seriously since July 2013, when President Barack Obama issued Executive Order 13648 on Combating Wildlife Trafficking. The order states: "Wildlife trafficking reduces those benefits while generating billions of dollars in illicit revenues each year, contributing to the illegal economy, fuelling instability, and undermining security; . . . it is in the national interest of the United States to combat wildlife trafficking."[59] Indeed, one Washington, D.C.-based conservationist pointed out that "the State Department under Obama was really dedicated to cracking down on the illegal ivory trade. One of the undersecretaries of state made it her personal mission."[60] Further, from the perspective of a USFWS official, their office had worked on wildlife trafficking issues for twenty-five to thirty years, but "in the past five years, doubly since 2013, with the executive order from Obama we saw this huge upswing in interest, so that is definitely a change but we are still a tiny office but there is much bigger interest in what we do from multiple departments. For example, the State Department, 10 years ago they would not have imagined talking about wildlife conservation. . . . Here having a presidential executive order gave everyone license to bring it up at a high level. If you are in an embassy you can say, look this is what our president said, this is important. It came up at the G20."[61]

There were concerns that such high-profile support from the US would evaporate with a change of president. At the start of the Trump presidency in 2017, conservation organizations were worried that illegal wildlife trade might not remain a government priority. The END Wildlife Trafficking Act[62] was passed in 2016, at the end of Obama administration, and there were fears that the Trump administration would not implement it. Therefore, once Trump entered the White House several conservation organizations coordinated efforts to persuade the new administration to continue with support. The cross-party appeal of wildlife conservation was of great assistance. Wildlife conservation is a cross-party issue in the US government. It is also supported by the powerful bipartisan International Conservation Caucus (ICC).[63] As one Washington, D.C.-based conservation lobbyist remarked, "The nice thing about this is, and its rare in the US, wildlife trafficking is not a partisan issue so we actually have, after Trump was elected, [name] and I were calling a bunch of supporters. . . . Every single lawmaker we talked to, Republican and Democrat, reconfirmed their belief in the importance of the END Wildlife Trafficking Act. So at least it is still is on their radar."[64] When the Trump Administration specifically highlighted wildlife trafficking as part of the Executive Order on Transnational Crime in 2017, it was regarded as an indicator that the President did take it seriously.

The militarization of conservation is one outcome of the intense interest in the security implications of the illegal wildlife trade, but it could be argued that in some instances it has gone a step further to become *war by conservation* (see chapters 4 and 7).[65] In brief, it denotes the ways in which conservation has become part of a broader strategy of the global War on Terror. It can be used to develop, try out, and refine new military techniques. One of the main drivers is security and stabilization of areas that are of geostrategic interest to the US-led War on Terror. Conservation is then mobilized as a new means by which preexisting US national security concerns can be addressed. It is not just a reactive turn back to the barriers or fortress conservation movement, anchored in a forceful defense of clearly defined protected areas. Instead, it can be characterized as an offensive position in which conservation is an aggressor, not just the defender, of wildlife. This new phase of war by conservation differs from militarization precisely because conservation agencies, and even NGOs, are becoming engaged in the use of force against people they identify not only as poachers but as members of armed groups or terrorist networks.

MILITARIZATION TO PROFESSIONALIZATION AND LAW ENFORCEMENT

At first militarization was rolled out as the solution to an urgent problem. Such pressures can be especially acute in conflict zones or in areas where conservationists have identified poachers as engaged in more aggressive tactics.[66] It was also clearly intensely focused around specific high-profile, charismatic, and high-value species. As one Washington, D.C.-based conservationist commented, "Militarization seems to me to be commensurate with the popularity of the species. You hardly hear about armies of heavily armed rangers protecting pangolins, even though they are much more threatened. . . . Pangolins are really up against the wall and you don't see this kind of swat team type approach to conservation. . . . If somebody were to run the numbers on this, they would probably find a direct correlation between the popularity of the species or the familiarity of the general publics in Europe and North America with the militarization, which suggests there is a connection."[67]

Militarization relies on enforcement of existing protected areas and regulations. Several academic studies have criticized militarization of conservation[68] in a wide range of cases, including Laos, Guatemala, Colombia, Democratic Republic of the Congo (DRC), Nigeria, India, and Tanzania.[69] Militarization does not attempt to address or redress the injustices produced by the colonial histories of conservation, especially in Africa, and it may even exacerbate them.[70] The focus on detection, interception, arrest, prosecution, and even killing of poachers concentrates on the symptoms instead of on the much deeper structural drivers of illegal hunting. For instance, militarization does not address or tackle inequality, poverty, or lack of opportunity, nor does it address historical grievances about eviction and displacement for wildlife conservation initiatives, including protected areas.[71]

The arguments in favor of militarization also overlook the ways that local communities experience this form of conservation practice. More militarized approaches have the capacity to alienate local communities, losing their support in the longer term.[72] In some cases, local communities do welcome greater enforcement of protected areas in order to provide greater security. In many more cases, however, the experienced militarization of conservation is a return to or extension of forceful and oppressive state practices. For example, in South Africa the extension of militarization means that, from a local community point of view, current conservation practice can appear very similar to the apartheid-era practices of raids on homes, beatings, arrests, and intimidation.[73] These more forceful strategies are accompanied by schemes to win hearts and minds by providing incentives, such as better water supplies, game meat distribution, and education programs.[74] One national park manager, who was operating in a context of very limited state capacity, high levels of poverty, and protracted armed conflict pointed to the importance of integrating conservation with prospects for development, told me: "We have been funded by [donor] for what is called the peripheral, development of the periphery of the park. It's not a very good term for it, it's socioeconomic development linked to conservation, so that there is a belt around the park that receives direct economic benefit because of the park. This could be through hydro . . . infrastructure improvements . . . agro forestry, it could be protein source."[75]

Another area often overlooked is the impact of militarization on conservation staff. As Annecke and Masubelele point out, many rangers joined conservation services before the development of more militarized approaches, and they were not expecting to work in such conditions.[76] As a result, there are growing reports of rangers suffering from stress, exhaustion, and even PTSD. On its website, the Game Rangers Association of Africa notes: "Rangers are expected to go beyond their typical role as conservationists to become active players in guerrilla warfare, putting their lives in constant jeopardy. Rising incidents of PTSD, acute stress disorder and burnout fatigue are just some of the effects rangers have to endure as a result of this on-going assault on our natural heritage."[77]

Greater levels of workplace stress and exhaustion are often attributed to the extension of the working day and changes of working practices. One of the consequences of the introduction of new monitoring technologies is that it can produce a culture of being constantly on call to respond to suspected poaching incidents.[78] WWF recognized the need to develop a better understanding of how rangers are responding to the pressures of this new phase in conservation practice. In 2016, it launched the *Ranger Perceptions Survey: Asia*[79] and *Ranger Perceptions Survey: Africa,*[80] which revealed ranger concerns about low pay, job insecurity, and poor equipment.

It is also important to note that militarization can present new opportunities for conservation workers to develop their professional careers, as Lombard and Tubiana demonstrate in their in-depth analysis of three individuals working in conservation in eastern Chad (primarily in and around Zakouma National Park) and northeastern Central African Republic (primarily in and around Bamingui-Bangoran and Manovo-Gounda St. Floris National Parks). Their biographical approach clearly shows how militarization can also be understood in terms of the individual agency and career ambitions of tracker guards; each tracker guard sought to find a place in broader markets for arms-carrying work, which provided social status and employment stability. Put simply, armed anti-poaching work was woven into longer, and more wandering, professional careers and personal lives.[81] Dutta also remarks on similar dynamics on the creation of the Ecological Task Force (ETF) in India, which attracted recruits who were former military; joining the force was the latest step in a longer timeline of career progression.[82] Professionalization and training provides important opportunities for conservation staff, in terms of building capacity and confidence especially in areas marked by armed conflict. As one protected area manager, operating in a conflict zone, stated, "[Rangers] join and they do the same job, some of them for up to thirty years. I don't think it is particularly healthy. But there isn't much of an alternative if you are from the local area, which many are; standards of education are fairly low, birth rates are high, dependents are numerous, so you could be providing the main wage going into quite an extended family. . . . There is the wage, housing, child benefits, school, medical; we are improving so that we pay them on time. And it's a fairly good wage for where we are and we are aiming to improve the welfare side with additional facilities."[83]

Militarized forms of conservation are often presented as essential and as the last resort for saving wildlife in areas of intense armed conflict. At first glance, it seems a reasonable case to make when conservation projects have to operate in places where armed groups are active. As one representative of a major donor organization remarked, 'the stabilization benefits of well managed parks are obvious. . . . [National Park has] become an enormous economic project."[84] For example, African Parks, which operates in a context of wider conflict in Chad and CAR, argues that conservation can be an effective form of counterinsurgency. Armed forms of conservation, they suggest, protect wildlife and the people living nearby; further it can provide development and services, replacing the absent state security forces or civilian administration.[85] There are similar arguments about the role of African Parks Foundation as a conservation authority in Garamba National Park in DRC as the foundation also provides services and development to local communities and operates in the challenging context of protracted conflict and absence of state capacity.[86]

Indeed, one security specialist working with national park authorities in an area of intense regional conflict remarked to me that wider security was "definitely important because it means we can't look at [National Park] as a simple protected area management scheme; we can't look at it in the same ways as a South African ranch, or reserve, park, or protected area."[87] Another protected area manager I spoke with, one who was working in a region characterized by protracted long-term conflict, including cross-border raids, was keen to draw attention to the hidden costs of attempting to engage in conservation practice in areas of conflict: "Because we have to have quite a big law enforcement insurance policy, which bloats the budget for the park and makes the operating costs high, and when you are an NGO getting that funded is a challenge. So we do have a preoccupation with security, and we are trying to go about it in a sustainable way, keep the costs down but in a sustainable way."[88]

However, a missing part of the puzzle is that conservation initiatives are rarely equipped to address the underlying drivers of the conflict, which threatens the wildlife. In fact, conservation initiatives can contribute to deepening the conflict.[89] Rangers are not a singular category of heroic individuals who work with wildlife because they love nature. The reality on the ground is that there are diverse ranger experiences, concerns, and activities, including implication in human rights abuses.[90] Further, there are questions surrounding accountability in conservation schemes that operate in places of protracted conflict and wider militarization of society. In examples like Virunga, the park authorities and their international supporters can constitute important sources of economic development, security and stability. But power dynamics inevitably arise between state authorities, international donors, and conservationists.

The criticisms of militarization of conservation are not confined to academic studies. Some conservation practitioners have spoken out about these approaches to anti-poaching, especially in places where they are not combined with efforts to address local-level socioeconomic inequalities and historical injustices (such as the exclusion and dispossession of communities to make way for protected areas).[91]

In the case of Kruger National Park, Lunstrum argues that militarized responses are seen even by those leading them as short-term solutions to *stop the bleeding.*[92] As one conservationist based in Washington, D.C., suggested to me: "There are pockets of resistance to it [militarization] everywhere. But it's difficult to argue with when doing it right takes much longer. . . . It's a race against time, and I think that a lot of caution is being thrown to the wind in an effort to save the species."[93] Further, there was a warning that militarization of conservation could have long-term

South African National Defence Force camp set up in Kruger National Park, along the border with Mozambique. Photo credit: Francis Massé.

negative effects because the "bill will come due later and we are definitely not prepared to deal with it. . . . The short-term solution is militarization but it is going to spawn, it will perpetuate a lot of inequalities."[94]

The criticisms of militarized conservation offered by academic researchers and conservation practitioners were echoed in criticisms by NGOs with a focus on human and indigenous rights. One US government official pointed out that "there are groups which are pushing back on green militarization . . . [because] in many cases this can turn to further disenfranchisement of local communities and human rights abuses if there is an overly militarized approach."[95] NGOs campaigning on indigenous and human rights have been critical of militarization. Since 2015, Survival International has run a campaign entitled Stop the Con about the abuses of local communities around Kaziranga National Park in

India. Similarly, it draws attention to the abuses perpetrated by eco-guards against Baka and BaYaka forest communities in Cameroon and in Central African Republic respectively.[96] Survival International has sought to highlight the ways indigenous and tribal peoples are able to manage and conserve natural resources far better than the big conservation NGOs. They state that 80 percent of the world's most biodiverse areas are home to indigenous and tribal peoples.[97] This runs against the assumptions, often made in conservation circles, that conservation is best served by separating out people and wildlife. In a well-documented case of abuse of Baka peoples in Cameroon, the eco-guards responsible were trained by a private company, which was ultimately funded by WWF. Survival International states: "In Cameroon, Baka tribesmen who dare to enter the forest they have been excluded from are terrorized by anti-poaching squads funded by WWF. In India, tribal villages are expelled from tiger reserves at the same time as the forest department encourages rocketing tourism. The big conservation organizations are complicit. They fund militarized conservation which leads to the persecution of innocent hunter-gatherers, they partner with the big businesses that steal tribal lands, and they drive the projects that result in illegal evictions."[98]

In March 2019, Buzzfeed News published an article based on a year-long investigation into human rights abuses by rangers employed or trained by WWF. They uncovered beatings, killings, and instances of torture in Chitwan National Park (Nepal), Kaziranga National Park (India), and across several places in Africa. It included the abuses of the Baka communities that Survival International had already highlighted. The investigation caused some outcry, and WWF-International hired a legal firm to undertake an internal investigation into the allegations.[99] The US Government Accountability Office report confirmed that WWF was aware of the abuses by the eco-guards they had supported.[100] In November 2020 WWF-International released its report on the internal investigation into the allegations of abuses. The report stated that the main finding of the expert panel was that there was "no evidence" of WWF staff directing or engaging in abuses.

However, the report amply showed that WWF staff were aware of abuses, failed to act, and continued to fund those initiatives.[101] Rainforest Foundation UK (RFUK) criticized the report as narrowly focused on a few specific allegations in the Buzzfeed News report, despite evidence of wider abuses, especially at Salonga National Park in DRC, where RFUK had already drawn attention to instances of torture and rape by park guards.[102]

RFUK also produced several reports on abuses of forest peoples as a result of wildlife conservation initiatives in the Congo Basin.[103] For example, Baka, Aka, Bagyeli, Bakola, and Batwa peoples in the Congo Basin have traditionally engaged in hunting and fishing in protected areas to meet their protein needs.[104] Wildlife is critically important for their day-to-day survival, yet increasing levels of (often violent) enforcement of park boundaries and regulations have effectively prevented them from engaging in the hunting they rely on. As a result, there are reports of malnutrition in some communities.[105]

As these criticisms have grown, and militarized conservation practices developed on the ground, there has been a discernible shift in language and practice. More recently, conservation has started to emphasize tackling wildlife poaching and trafficking as a crime and a matter for policing (see chapter 2). This has been accompanied by a greater emphasis on separating out different aspects of militarization. Examples include training, professionalization, and capacity building, rather than heavily arming rangers and endorsing shoot-to-kill policies.

As the arguments about the merits and disadvantages of militarization developed, the term "militarization" became more and more associated with violent approaches, obscuring the ways that military-style thinking, tactics, and training can contribute to wider forms of militarization. As a consequence, some in the conservation community moved away from using the term. One conservationist, working in a conflict zone, clearly captured the complex and messy entanglements involved in partnering with military actors: "I felt I was very wary of these very heavy-handed, clumsy analyses of the situation, very narrow understanding of the complexity context and adapting to it. So, I felt a bit indignant that I was being lumped with shoot-to-kill. I already have to defend my position because I was kind of feeling misunderstood. . . . It all depends how it is done."[106] These sentiments were echoed by another conservationist running an anti-poaching project in a conflict zone, who suggested: "Militarization isn't one thing, it is a tool, how it is used. And I think that is the key thing for conservation. You have got a suite of tools and you have to adapt them to the situation."[107]

The growing emphasis on professionalization and training of conservationists, especially of rangers, contributes to militarization. Those trained in conservation are inspired by military approaches to understanding the world, especially when thinking about problem solving, risk reduction, and a narrowly defined form of security. For critics, many of the very same practices that characterized militarization were now described in more neutral or positive terms, such as law enforcement, professionalization, and tackling wildlife crime. This shift in language serves to obscure the range of forceful and militarized practices that are subsumed under the idea that professionalized staff are engaged in upholding the rule of law.

The militarization of conservation has developed and changed over time. It is not a singular thing but instead encompasses a range of approaches and practices. Different forms of militarization operate in very different contexts, ranging from contexts of stability and strong state capacity to areas of protracted conflict, weak governance, and an absence of state presence or capacity. It is important to draw out these nuances in approaches, practices, and operating contexts, since each shape the precise form that militarization takes. The shifts toward more forceful forms of conservation are not necessarily new. They build on the militarized origins of some of the first national parks and the long-standing model of fortress conservation, as well as on the poaching crisis of the 1970s and 1980s. Furthermore, the militarization of conservation builds on the earlier neoliberal phase, drawing more and more aspects of nature into market logics. For example, the current security phase, of which militarization is one part, has allowed new players to enter conservation. This has been supported by the increasing levels of funding available from donors, NGOs, governments, and the private sector, all seeking solutions to the poaching crisis. The new streams of funding have encouraged and developed new markets for security-related products and services. The donor and philanthropic community have provided funding for more forceful forms of conservation with the promise that they can deliver and meet targets. It is much more difficult to argue that long-term community-based work or demand-reduction programs can meet targets in the short time frames that donors and philanthropists often prefer.

Military-style training can provide rangers with skills in deescalation and restraint; militarizing conservation in conflict zones can be an understandable response under intense pressure because it can provide short-term protection for staff and wildlife. However, it does little to tackle a central contradiction in current debates about illegal wildlife trade: the failure to engage with the dynamics of global political economy as underlying drivers, or causes, of unsustainable and illegal trading in wildlife products. Further, it does not address the underlying reasons for wider forms of armed conflict, which also shape conservation in some areas. Without addressing these issues, militarizing conservation will just add to the number of armed actors and a militarization of wider society.

Furthermore, militarization can deepen and extend existing power dynamics in conservation, which can actually sustain the illegal trade. Security-style approaches can result in supporting and maintaining the position of social, political, and economic elites. The international networks of conservation NGOs, donors, philanthropists, government wildlife agencies, and private companies may engage in conversations about the best way forward for tackling the illegal wildlife trade, but one voice is often missing: that of communities and marginalized peoples living with wildlife. As a result, marginalized peoples are unable to shape and drive conservation policy in directions that might be more sustainable and socially just. Militarization of conservation maintains and reenergizes a problematic model of separating of people and wildlife in strictly defined protected areas. In some places, militarized forms of conservation may be reaching a tipping point. This is the case when conservation becomes the means by which state agencies extend authority and control over people who are defined as resistant and unruly.

These events are fundamentally changing what conservation is, both philosophically and practically, on the ground. By aligning more closely with security thinking, including militarized techniques and practices, conservation has shifted in a direction that cuts out other options for the future. Ultimately, it is crucial that conservationists also reflect on the ripple effects of militarized strategies (including in areas of protracted armed conflict), especially when they produce human rights abuses and social injustices on the ground. Thinking in terms of what works best does not allow for important and challenging arguments about the wider implications of militarization. Moving away from a conversation about "what works best for now" opens up spaces for more sustained and careful thinking about how to tackle the illegal wildlife trade. It allows for thinking of ways to recognize and resolve the underlying drivers of illegal wildlife trade: global political economy, inequality, wealth, poverty, and historical injustices.

4 TERRORISM AND POACHING

There have been high-profile claims that the illegal wildlife trade funds terrorism, and these claims have moved it up the political agenda, increased funding, and been used to justify the need for a militarized response on the ground. A key moment occurred in 2013, when claims emerged that ivory was a central source of funding for Al-Shabaab, a group that has allied itself with Al Qaeda and is defined as an international terrorist network by the United States government. This claim, that ivory funds terrorism, originated in a 2012 report by the Elephant Action League and was later taken up widely by the media, NGOs, and policy makers. The ivory–Al-Shabaab case is a good example of the ways that the illegal wildlife trade is presented as a security threat and how it has become a means of generating "threat finance." The argument that it is a form of threat finance takes two forms: it is a lucrative business for organized crime networks in Europe and Asia, or it is a source of finance for militias and terrorist networks (notably Al-Shabaab, Boko Haram, Lord's Resistance Army, and Janjaweed).

The argument that the illegal wildlife trade is a source of terrorist financing, and therefore is a global security threat, is central to the development of more security-oriented approaches in conservation. The key questions are how and why this dynamic has developed, who or what benefits from it, and, crucially, what kinds of blind spots are being generated by it? In this chapter my goal is not to prove or disprove the links between illegal wildlife trade and terrorism but instead to explore the political ecologies of these arguments and how they shape the current security turn in conservation. Peluso argues that, historically, the ultimate drivers of poaching have been overlooked in NGO campaigns because the focus was on the animals being poached, which were deemed to be the most important issue.[1] The tendency to overlook the drivers of poaching permeates the arguments that poaching and trafficking are being used to fund armed groups. In drawing links between the illegal wildlife trade and terrorism, the conservation and security sectors have used what one US-based conservationist called "Washington dog whistles" that are guaranteed to gain the attention of powerful arms of the US state, including the Department of Defense and the national intelligence community. Drawing these links is changing conservation and restructuring the relationships between people and wildlife. It is a significant shift from a war for biodiversity, as discussed in chapter 3, to a new phase of *war by conservation,* in which initiatives to save species have been have been fused with global security efforts.[2] But how have these arguments shaped conservation practice more widely?

GLOBAL SECURITY

A central component of the growing global interest in illegal wildlife trade is the argument that it is used to fund terrorism, and therefore it constitutes a global security threat. This argument has gained attention and attracted funding from a range of actors, including conservation NGOs, donors, and private military companies. It has also grabbed the attention of the international news media, social media platforms, high-profile filmmakers, and it has captured the public imagination. Intuitively it makes sense that terrorist networks use the lucrative illegal trade in ivory (especially) as a source of income. However, there is a question mark over whether such an intuitively acceptable story is actually credible. The story of how this particular set of arguments rose to prominence and the influence they have had on thinking about conservation and about security is both fascinating and instructive. The story that ivory funds terrorism fits very well with the security concerns of some of the world's most powerful actors. As such, it is a key reason why more security-oriented approaches to conservation attracted new sources of funding: they offered a triple win—save wildlife, fight terrorism, and enhance security.

The argument that the illegal wildlife trade is a global security threat takes three distinct forms. First, that poaching and trafficking undermine the rule of law, encourage corruption, and contribute to instability in governance. A good example is provided by the EU-funded European Union Action to Fight Environmental Crime (EFFACE) project case study report on illegal wildlife trade in the United Kingdom, Norway, Colombia, and Brazil, which stated that "IWT contributes to civil conflict, economic loss, poverty, climate change, and negatively impacts on national security and stability, state authority and biodiversity and public health."[3] The second way in which poaching and trafficking are articulated as a global security threat is that armed groups use trafficking in wildlife products as a means of funding and extending their operations. For example, the International Fund for Animal Welfare (IFAW) report "Criminal Nature" made an explicit claim that the trade provides funds for rebel groups, militias, and terrorist organizations.[4] Similarly, Born Free USA commissioned a report by C4ADS (an organization with a specialism in conflict and security) into the ivory trade, which claimed that ivory has funded conflicts across Africa. It specifically identified Al-Shabaab in Somalia and Boko Haram in Cameroon as engaged in the illegal ivory trade.[5] The third form revolves around the *convergence* of wildlife crime with other illicit trades as a source of funding for activities that constitute security threats. There were concerns about the ways wildlife trafficking was mixed in with the trade in heroin from Mozambique, through South Africa to Europe.[6] Criminal networks are not necessarily specialists in one illegal trade, but rather they opportunistically switch between trafficking vehicles, weapons, people, drugs, counterfeit goods, and wildlife.[7] A 2019 European Commission report also highlighted the links between wildlife and security in Africa, stating (in relation to South Sudan) that "wildlife crime helps drive insecurity with armed rebel groups surviving in remote areas and living off wildlife (bush meat). Groups and individuals engaged in poaching are often also involved in other activities that create insecurity, such as banditry, robbery and child abduction. The violence and pillaging destabilizes local communities and undermines development opportunities."[8]

There are long-standing debates in a wide range of academic fields, including security studies, peace and conflict studies, politics, international relations, and human geography about the difficulties of defining terrorism and terrorists.[9] But these debates are not well reflected in arguments about the illegal wildlife trade and terrorism. Poaching and wildlife trafficking are being redefined as a global security threat by a complex mix of journalists, NGOs, national governments, international organizations, and private companies.[10] For example, a policy brief by the EFFACE project provides a good example of this shift in thinking. It recommends "encouraging UN troops to support wildlife rangers in source countries where organized crime groups operate."[11] Similarly, the European Commission has also highlighted the need to integrate conservation with security. A 2019 report on the intersections between wildlife conservation in Africa makes the following recommendation: "The complexities of the subject and the links between wildlife trafficking and conflict, insecurity, the impact on socio-economic security, the rule of law and migration examined here demonstrate the necessity to expand and increase investments in conservation-security-development programs in priority protected area landscapes of sub-Saharan Africa in order to achieve sustained global, regional and local security and stabilization objectives."[12]

Many conservation NGOs, donors, and government agencies refer to terrorism and global security in campaigns about the illegal wildlife trade. As one senior UK civil servant commented: "Hopefully if we move it [IWT] up the agenda of political attention, that is partly about collecting data and pulling it together. If people aren't looking for it then they won't spot it. . . . I am absolutely convinced it is the right way to talk about it, it is really helpful, but security means different things to different people."[13]

Gaining public attention is a key issue facing the policy networks involved in tackling the illegal wildlife trade: how do they talk about it in ways that will attract the attention of senior government officials, presidents, the general public, and donors? It seems an obvious strategic move to link the illegal wildlife trade to the central concerns of governments, which revolve around security and strong economic performance. A senior US government official suggested that the illegal wildlife trade can flourish "when countries are lawless and not really well monitored then that is the cradle of destabilizing. It can be considered jihad. . . . They function best in very unstable, and very disorganized, and ungoverned spaces, so their actions I believe are deliberately oriented to maintaining the anarchy in which they can thrive."[14]

Yet it is not so simple: making conservation a security problem prompts a security-type response, which may not tackle the underlying drivers of poaching and trafficking in the first place. Therefore, the ways that wildlife trafficking and terrorism are linked in debates about conservation can have far-reaching implications.

POACHING AND GLOBAL TERRORISM

Conservation NGOs have argued that the illegal wildlife trade is a source of funding for international terrorism. In considering this possible connection, the first question that arises is how do we define terrorism? What are the challenges in identifying a particular group as a terrorist organization, and how do we define terrorist financing? As Tim Wittig's study of terrorist financing points out, we actually have a very rudimentary understanding of how it operates, and much research and policy activity has focused on the very narrow question of who finances terrorism and how do they do it.[15] There is an assumption that terrorist networks operate in a shadowy world of illicit financial activity. However, it is more accurate to think of terrorist financing as a fluid structure in which terrorist networks regularly interact with formal and legal sociopolitical entities, as well as illicit and secretive worlds, in order to raise funds.

To date the main focus has been on how poaching and wildlife trafficking might support armed groups and terrorism networks based in sub-Saharan Africa. There have also been claims of poaching and trafficking by groups in India, Bangladesh, Thailand, Nepal, Indonesia that are linked in to Al Qaeda, as well as by jihadist groups in Mali and Seleka and anti-Balaka movements in Central African Republic.[16] But the main claims about terrorism revolve around stories about wildlife poaching and trafficking in sub-Saharan Africa by Al-Shabaab, Janjaweed, and Lord's Resistance Army and, more recently, headlines about Boko Haram. It is important to understand these fully because they are central to justifications for integrating conservation and security.

First a note on evidence. It could be argued that the evidence base for links between poaching and terrorism is held by organizations such as INTERPOL or the CIA, and is therefore classified or confidential, and if such evidence is held then it is not made public. In various discussions and meetings I attended while researching this book, the existence of such evidence was referred to but never presented on the grounds that it was highly sensitive, confidential, and covered by concerns about national security. Therefore, it is not possible to scrutinize this information and make judgements about its credibility. Rather, we can only examine what is available in the public domain, and that evidence base is very narrow.

The arguments about terrorism and illegal wildlife trade have been taken up in global debates about how best to tackle poaching and trafficking for two reasons. First, because they tap into a preexisting and deep-seated global fear about the expansion of terrorism networks post 9/11. Second, because arguing that conservation can contribute to security might offer the potential for a new and lucrative stream of funding. The enhanced levels of funding available for tackling the illegal wildlife trade from a security perspective has supported and facilitated the turn toward more security-oriented approaches in conservation.

It is useful to give some specific examples that illustrate how the links are drawn. In 2014 one of the world's largest and most prominent conservation NGOs, Conservation International, clearly stated that it believed "money from wildlife poaching and trafficking is directly linked to the funding of dangerous rebel organizations and terrorist networks. These include the Janjaweed militia in Darfur, the Lord's Resistance Army in Uganda, and Al-Shabaab in Somalia, which is now linked to al Qaeda."[17] Conservation International also argued that conservation has a direct link to US national interests, especially economic interests and security; the "Direct Connection" campaign stated that competition over scarce resources can lead to conflict, instability, and failed states. These help terrorist networks recruit members, cut off trading opportunities for American companies, and produce large-scale illegal migrations to the US.[18] This scenario reflects the idea that since poachers pose a clear and present threat to global stability, forceful action against them is justifiable. In his public statements Peter Seligmann, CEO and chairman of Conservation International, has made the connection between poverty, trafficking, and threats to global stability. For example, commenting on the new Clinton Global Initiative support to end wildlife trafficking, he emphasized that "what we're seeing here is the perfect storm of extinction, poverty, and radicalism. We're seeing the deterioration of societies and a massive threat to the stability of not only African nations but the entire world. A crucial step in changing this equation is to ensure that the ivory trade comes to an end."[19]

Conservation International produced a short film, *Direct Connection,* using Harrison Ford (actor), Wes Busch of Northrop Grumman (director of a global security company), and Rob Walden (chairman of Walmart Stores, a global corporation best known for its department stores), which, as its title suggests, underlines the link between conservation and US national and economic security.[20]

The argument that the illegal wildlife trade contributes to global instability has captured the imagination of politicians, conservationists, and the wider public. It is a story that has gained traction because it brings together much-loved wildlife, a sense of urgency, and fears of extinction with much wider anxieties about new kinds of networked global security threats, most notably Islamic extremism. But there is a risk that such links between poaching and terrorism achieve the status of truth or fact by repetition, rather than through the presentation of evidence.

Given the stakes involved, it is worth examining the evidence base provided by Conservation International for its public statements. For example, in his statement, Peter Seligmann cited a blog from Slate.com, which in turn referenced the investigation by Elephant Action League (EAL) into ivory trafficking and the Westgate Mall attacks in Nairobi by Al-Shabaab.[21] The nature of EAL's investigation is very important to the ways poachers are described as terrorists.

Conservation International is not alone in this. The 96 Elephants Campaign by Wildlife Conservation Society, another leading NGO, was originally organized around three central pillars: Humans and Elephants, Terror and Ivory, and Heroes and Hope. This campaign linked poverty, regional instability, poaching, and terrorism.[22] Under the topic of Terror and Ivory the campaign made a series of statements but did not provide references to support its claims. Instead, it quoted public statements by Hillary Clinton and by Congressman Ed Royce (co-chair of the International Conservation Caucus of the US Congress). The 96 Elephants Campaign referred to ivory as the "white gold of Jihad," a reference to the terminology of the EAL report cited by Conservation International.[23]

AL-SHABAAB, IVORY, AND SECURITY THREATS

The idea that ivory trafficking funds terrorism has had immense sticking power, and it has been used to justify much more security-oriented approaches to conservation. But what are the main features of the ivory-terrorism debate, why did it rise to prominence, and how is it sustained? Andrea Crosta and Nir Kalron of Elephant Action League played a central role in creating and spreading the message that Al-Shabaab generated significant revenue from ivory trafficking. The messages from their report *White Gold: al-Shabaab and Conflict Ivory,*[24] has been enthusiastically taken up and repeated by a range of global actors; it has shaped global-level policy shifts since 2013 precisely because it mirrors and supports existing fears about Al-Shabaab, regional instability, and global security.

The year 2013 is significant; the narrative of *poachers as terrorists* gained traction following the Westgate Mall attack in Nairobi, 21–24 September 2013.[25] The *White Gold* report was uploaded to the EAL website in 2012 but was reported by the international media only after the attacks on the Westgate Mall. Since then, the argument that Al-Shabaab uses ivory to raise funds has become commonplace. It has been repeated in several news media reports, NGO reports and statements, and a range of published documents. Each of these are then offered, in further reports and statements, as a wider evidence base for the ivory-terrorism argument, despite the fact that they all rely on the same single source. These further mentions include media reports in national newspapers, such as the UK's *Independent* newspaper,[26] NGO campaigns by IFAW and Born Free Foundation, and a report by the UK's Chatham House.[27] For example, in 2013 the IFAW report *Criminal Nature: The Global Security Implications of the Illegal Wildlife Trade* devoted a whole chapter to "Illegal Wildlife Trade Links to Violence, Radicalism, and Terror."[28] The authors of that report repeat the allegations that Al-Shabaab is engaged in ivory poaching and trading to fund its operations. But the only evidence offered is statements made by other NGOs (such as Ian Saunders of Tsavo Trust) and news media reports from the *New York Times, Vanity Fair,* and *Spiegel Online International.*[29] The claims in the EAL report *Africa's White Gold of Jihad* were also reported uncritically by news media outlets around the world, including the *Los Angeles Times,*[30] *National Geographic,*[31] *Voice of America,* and the UK's *Independent* newspaper.[32] Following the Westgate Mall attacks, then-President Kenyatta of Kenya wrote a piece for the *Wall Street Journal* in which he stated that Al-Shabaab acted as facilitators and brokers in the ivory trade.[33] Such arguments fit well with contemporary global fears about international terrorism and with the interests of national governments attempting to tackle such movements within their own borders.

These arguments have informed high-profile initiatives aimed at the general public, including short animated films posted on social media. For example, in December 2014, Oscar-winning film director Kathryn Bigelow released a short animated film, *The Last Days of Ivory,* in conjunction with WildAid, which directly links buying ivory in China to funding conflict and terrorism in the Horn of Africa.[34] The film's campaign slogan, "End Ivory Funded Terrorism," is intended to be a fundraising vehicle aimed at English-speaking audiences. It is possible that the reliance on the EAL report—and later news media coverage—as the evidence base partly accounts for the remarkable similarity of the statements on ivory, terrorism, and Al-Shabaab.

Since the claims were so widely reported and shape what types of conservation is funded by donors, governments, and the private sector, they merit greater scrutiny. The report was based on eighteen months of undercover research in Somalia, where EAL researchers were able to interview one individual from within Al-Shabaab who claimed that there was one trader on the coast who occasionally traded ivory and that the ivory sometimes came from Al-Shabaab operatives. The video and audio evidence remain confidential due to fears about reprisals against informants. The evidence was shown in confidence to a number of security agencies around the world.[35] However, it is not clear what role the person identified by the trader actually played in Al-Shabaab, whether they were high ranking, deeply involved, or simply linked in some way.[36] Further, Somerville points out the report also contains lots of inconsequential detail and questionable claims and is written in the style of a spy thriller, likely to maintain readers' attention. The result is a narrative in which Al-Shabaab emerges as an important actor in a sophisticated network of traders in ivory and rhino horn.[37]

By contrast, a report by two researchers at the security think tank Royal United Services Institute (RUSI), Thomas McGuire and Cathy Haenlein, entitled *An Illusion of Complicity* offered detailed criticisms; they concluded that any Al-Shabaab involvement in the ivory trade to date is likely to have been opportunistic, ad hoc, and small scale.[38] The concerns around the links between ivory poaching and Al-Shabaab also reflected a more established and historical fear of Somalia as a source of instability and criminal activity. During the 1980s the KWS blamed Somali *shifta* (bandits) for crossing the border to wipe out the elephant population.[39]

Yet the story that ivory trading was central to Al-Shabaab's funding strategy gained traction, and it remains important and influential, even though it was largely debunked by 2015.[40] One US-based conservationist explained the sticking power of these stories as "one of the reasons it's pushed quite hard, about this link between terrorism and wildlife crime, is particularly ivory, particularly in the US . . . there is always an interest, and in other countries worldwide, if there is a link to terrorism then they will look at it."[41] Another also questioned whether ivory funded terrorism in any significant way at all: "I don't think the terrorist groups make a lot of money out of this. They might be able to exchange ivory for weapons. I don't know what a Kalashnikov costs or AK-47, but you might be able to get rid of one big tusk, and get two or three guns and buy bullets."[42]

Media outlets and campaigning NGOs tend to seize upon stories like ivory–Al-Shabaab and use the War on Terror framing and the international notoriety of these groups to draw attention to threats to elephants.[43] When such stories hit the headlines they help to increase attention and drive NGO fundraising efforts.[44] As one Washington, D.C.-based conservationist suggested, "The conservation community has not really expressed a whole lot of scepticism because it is driving money their way. This is a political economy of conservation. Bad news drives donations so there is some manipulation of the truth there too."[45]

However, the link is not just part of NGO campaigns; it quickly became the central legitimating argument of policy networks, especially in US government circles. It is less important in UK government circles, where civil servants in the relevant government departments were more sceptical and cautious about the claims.[46] In the US the story was very different indeed. One high-level representative from a donor organization commented that

> the interest from the US was that, for reasons which were never entirely clear to the outside world, a very strong focus on certain

> armed groups, or certain armed movements. Obviously, for security reasons, but possibly for economic reasons as well, and the main groups that they were concerned about were, obviously, Al-Shabaab in the Horn of Africa; Boko Haram in the, sort of, Cameroon, Chad, Nigeria, eastern Nigeria area; the Lord's Resistance Army, which is a different type of animal; and the ADF, which is a Ugandan-origin armed group which operates in northeastern DRC. And the American approach was basically a security-based approach, which was, let's throw Special Forces at these and see what happens. But it became more sophisticated, and by the end of the Obama administration I think all of us who are involved in these issues were having a dialogue which involved not just the military and intelligence, but also conservation, park management, civil society organizations, because in a lot of the places where these armed groups operate and where they do wildlife trafficking, basically there is no government.[47]

In 2012 the US Senate and US House of Representatives held a special congressional hearing, The Global Poaching Crisis. At the meeting witnesses carefully stated that ivory may fund Al-Shabaab operations or that ivory is an ideal commodity for groups like Al-Shabaab.[48] However, its overall headline conclusion was that "evidence is mounting that Al-Shabaab, an Al-Qaeda affiliate, and the Lord's Resistance Army are using these illegal animal products to fund their brutal campaigns of violence throughout the region."[49] At the meeting the founder of the influential US International Conservation Caucus Foundation (ICCF), David Barron, stated that "unless the United States takes strong action to combat the illegal poaching and trade of wildlife, terrorist groups will be increasingly fortified with funding and safe havens in Africa from which to launch attacks against the United States and our global interests."[50]

The ICCF drew the links between terrorism and ivory poaching in ways that intersected with the idea that underdevelopment and poverty lead to radicalism and instability; the ICCF stated that "ivory and rhino horn are gaining popularity as a source of income for some of Africa's most notorious armed groups, including Somalia's Al-Shabaab, the Lord's Resistance Army (LRA), and Darfur's Janjaweed. Illegal wildlife products are a substantial lifeline to African-based terrorism."[51] The supporting evidence for the claims are the statements of expert witnesses and web links to a 2012 article by Bryan Christy in *National Geographic* entitled "Blood Ivory, Ivory Worship"[52] and a 2012 *New York Times* article by Jeffrey Gettleman entitled "Elephants Dying in an Epic Frenzy as Ivory Fuels Wars and Profits."[53]

It is important here to delve into the statements made at the ICCF meeting by expert witnesses, all of whom are well-known individuals in the international conservation community. Expert witnesses are important because they act as important knowledge brokers when conditions are uncertain and risky, which characterize debates about the illegal wildlife trade. As such their opinions carry significant weight, they are listened to and taken seriously in policy-making circles. Furthermore, they can have an impact in shaping public understanding of poaching and its relevance to wider global security concerns.[54] They can impact particular understandings of poaching and the potential threats it might pose, enabling them to rise to international prominence via such platforms. For example, the 2012 hearing on the global poaching crisis heard expert witness testimonials from Ian J. Saunders of the Tsavo Trust, who claimed that rangers were engaged in low-level counterinsurgency against rebel groups. He stated that he believed "there is a credible, increasing security threat from Al-Shabaab in East Africa and that this will be fuelled from the wider illegal trade in ivory as long as the consumer states in the Far East continue to allow a domestic trade in ivory."[55] His statement made it clear that international action was required to save important species unable to defend themselves against an aggressor. By contrast, poachers and traffickers were presented as appropriate and legitimate targets for international interventions.

John E. Scanlon, then-secretary-general of CITES, also appeared as an expert witness to the US congressional hearing on ivory and insecurity. Regarding a high-profile ivory poaching incident in Cameroon, he said, "It was reported that elephants had been slaughtered by groups from Chad and the Sudan over several weeks, taking advantage of the dry season. The poached ivory is believed to be exchanged against money, weapons and ammunition to support conflicts in neighbouring countries."[56] In an interview with the *Guardian* (UK) newspaper a year later (in 2013) he stated that "the UN Security Council recently linked the Lord's Resistance Army to ivory smuggling in the Democratic Republic of the Congo, while Al-Qaida's Al-Shabaab group has been linked to illegal ivory in Somalia."[57] He could not have been clearer about how he thought ivory trading and terrorism were linked, but he did not state the evidence base on which he made his statements.

Further expert testimony at the US congressional hearing in 2012 was provided by Michael Fay, senior conservationist for Wildlife Conservation Society, who made similar supporting statements that tackling poaching would contribute to global stability and secure US national interests. For example, he told the hearing, "I would hold that bang for the buck, investments in the types of projects that I am involved in Gabon, Congo, Central African Republic, Tchad, Sudan, prove extremely productive not only for the cause of conservation, which I care deeply about, but to put out brush fires of illegal activity that degrade security in these nations, hurting US interests."[58] The influence of the hearings, expert witness testimonials, and NGO campaigns were discernible in the clear policy commitments to tackle wildlife trafficking and to list it as a serious crime during the Obama Administration.

While secretary of state (2009–13), Hillary Clinton endorsed the links between wildlife trafficking, poaching, and global security in her public statements, thereby lending the argument greater international weight. Her thinking was influenced by White House staff who were keen to get the illegal wildlife trade listed as a serious crime. Such a designation gave it a special status that facilitated information exchange and cooperation between US government agencies. Secretary of State Clinton was also influenced by media outlets and journalists working on ivory trafficking, notably by high-profile pieces in *National Geographic*. One journalist explained it as "after [Bryan Christy's] stories, it had a lot of waves of impact, he did a big media tour afterwards and that helped. . . . Hillary Clinton, when she was secretary of state, sent one of his [Christy's] ivory investigations to her colleagues; in her email, [she] said this is why the illegal wildlife trade needs to be on our radar. It seemed like the right people were reading it."[59]

The casting of ivory as "white gold of Jihad" (the phrasing used by EAL) was repeated several times in debates about the links between ivory and terrorism, including in a September 2014 op-ed in the *New York Times* by Monica Medina, a former special assistant in the US Department of Defense.[60] In the article Medina referred to a November 2012 panel, sponsored by WWF-US and *National Geographic,* on what the military could do to help, held in the run up to Hillary Clinton's announcement of a major State Department initiative to combat the illegal wildlife trade.[61] Hillary Clinton also appeared in the African Environmental Film Foundation (AEFF) film *White Gold,* in which she stated that terrorist organizations, including in Somalia, use ivory to fund their operations.[62]

Furthermore, the Clinton Global Initiative threw its weight behind the idea that wildlife trafficking constituted a security threat and required an urgent and robust response. For example, in 2013 the Clinton Global Initiative committed to raising US$80 million.[63] The funds were to be used for initiatives under three headline banners—"Stop the Killing, Stop the Trafficking, and Stop the Demand"—during 2013–16. The partners, or "commitment makers" as they termed themselves, included Wildlife Conservation Society, African Wildlife Foundation, Conservation International, International Fund for Animal Welfare, and World Wildlife Fund-US.[64] This again served to cement the idea that the illegal wildlife trade constitutes a global security issue and that it is a clear threat to US national interests.

However, since 2014 the ivory–Al-Shabaab link has come under intense scrutiny and growing criticism, to the point that the EAL had to release a statement that admitted its claims had indeed been overstated (but stopped short of accepting that there was no link). The basis of the EAL claims has been seriously questioned by several researchers and international organizations. For example, a report from UNEP and INTERPOL on environmental crime questions the accuracy of the links between ivory and Al-Shabaab. The report points out that ivory may be a major source of income for militia groups (especially Janjaweed) in DRC and Central African Republic, but, it notes that claims Al-Shabaab was trafficking 30.6 tonnes (33.73 tons) of ivory per annum (representing 3,600 elephants per year) through southern Somalia are highly unreliable and that the main sources of income for Al-Shabaab remain charcoal trading and financing from the Somali diaspora.[65] Indeed, as Menkhaus points out, if we regard Al-Shabaab as a criminal network, then their main criminal activities center on illegal charcoal trading, extortion, and protection rackets.[66] Put simply ivory is *not* a central pillar of their funding strategy.[67]

The argument of ivory as a key source of financing for terrorist networks and militia groups remains important. While there is now greater acceptance that the claims about Al-Shabaab and the ivory trade were poorly evidenced, they do form part of the argument about other possible Al-Shabaab links. For example, one senior UK civil servant commented, "There was one particular East African example [Al-Shabaab] that was pushed, and then it was said it's not evidenced, but that doesn't mean there isn't a link somewhere else, we need to keep looking for links whether it is terrorism or something else. It is not necessary to make the case that we should act."[68]

Such statements have been endorsed by more thorough reports by international organizations. A key example is the report by UNEP, CITES, IUCN, and TRAFFIC entitled *Elephants in the Dust,* which states that "political instability, armed militias, criminals, and most importantly, the rise in market demand, have once again resulted in a rise in poaching. . . . Poaching operations range from the old-fashioned camel and horse-based marauders to active intelligence units and helicopters, the use of which suggests substantial demand."[69] The report draws on a much fuller range of unpublished, confidential information, including data from the Monitoring Illegal Killing of Elephants (MIKE) and Elephant Trade Information System (ETIS) of CITES, as well as a number of academic studies on poaching rates. However, no source is provided for the statement quoted a few sentences above. Similarly, the European Commission produced a report on the intersections between security and illegal wildlife trade in Africa, which states, "Wildlife poaching and trafficking are driving conflict and insecurity in areas where heavily armed poaching gangs are crossing borders, killing elephants and other wildlife, and creating insecurity for local communities living in and near protected areas. There might be terrorist group involvement in ivory poaching and trafficking in vulnerable areas of sub-Saharan Africa, although at present few cases have been confirmed."[70]

It is really important to examine the evidence base for the high-profile claim that trafficking ivory funds terrorism. What is clear is that there is very little publicly available evidence for it. I have questioned the evidence base in the course of many conversations while writing this book, and I have been greeted with one of three reactions: there is agreement that there is little or no evidence for the claims; there is a link via repetition because *everyone knows it*; or, finally, the evidence does exist but it is classified. Whatever the truth is about the link between ivory and terrorism the resilience of this argument has shaped and driven conservation policy. It has encouraged and facilitated security-style approaches to tackle the illegal wildlife trade, and it has provided a rationale for more militarized forms of conservation. But if the ivory–Al-Shabaab link is poorly evidenced, then what is the evidence base for claims about poaching by other armed groups?

LRA, JANJAWEED, BOKO HARAM, AND WILDLIFE TRAFFICKING

There are high-profile claims about the involvement of armed groups in poaching as both a conflict strategy and source of finance. These include the LRA, Janjaweed, and most recently Boko Haram. A common thread is that these three groups are central to European and American popular fears about sub-Saharan Africa as a source of instability, that they thrive on destabilization and as such they constitute global security threats.

IFAWs *Criminal Nature* report states that the LRA and Janjaweed are engaged in significant levels of elephant and rhino poaching. But the sources cited in the report are news media reports and a statement from Emmanuel de Merode, director of the Virunga National Park, where none of these armed groups actually operate.[71] A fuller and more sustained investigation by UNEP/INTERPOL indicates that ivory *may* be a major source of income for militia groups (especially Janjaweed) in DRC and the Central African Republic.[72] As a final example, the European Commission report on the intersections between illegal wildlife trade and security in Africa identifies LRA as a terrorist group involved in the ivory trade; it also points out that ivory poaching and trafficking in and around Garamba National Park is carried out by Sudan People's Liberation Army (SPLA), Sudan People's Liberation Movement (SPLM), and associated splinter groups, as well as by elements of DRC national army, the Forces Armées de la République Démocratique du Congo (FARDC).[73] Three of my interviewees with experience on the ground in South Sudan and DRC agreed that poaching was carried out by this range of groups operating in the area, and they confirmed that while LRA was involved in an ad hoc way, it was not a major player.[74]

The ways that poachers are linked to organized criminal and terrorist activity is illustrated by United for Wildlife (UfW), an initiative by the Duke and Duchess of Cambridge and Prince Harry via the Royal Foundation. It brings together leading conservation organizations[75] to cooperate in responding to the apparent rise in poaching and trafficking. It states that "alleged connections to armed groups such as Lord Resistance Army in Uganda has led to ivory being dubbed *Blood Ivory,* bringing a human element to this already tragic story. By reducing demand, the funding that these groups gain from their criminal activities will be directly affected."[76] UfW's #whosesideareyouon campaign encourages supporters to choose sides "between wildlife and the criminals who kill them for money." UfW has a powerful position and role in shaping public debate around conservation and the illegal wildlife trade.

UfW arose out of a confluence of different factors which are indicative of how influential the British royal family has been in shaping public debates and government policy toward illegal wildlife trade. For example, it could be argued that the UK government response, including its four high-profile conferences on tackling the illegal wildlife trade, were kick-started by an influential intervention by Prince Charles. He convened a high-level meeting at St. James Palace in May 2013 to discuss how the UK should respond to the rises in poaching in sub-Saharan Africa.[77] After that meeting there were a series of government-level meetings to determine the policy response, and in December 2013 the UK government announced a £10-million fund to combat trafficking. The fund provides support for practical steps to combat trafficking and poaching.[78]

There are, however, reasons to doubt that LRA and Janjaweed are engaged in significant levels of poaching and ivory trafficking. Titeca argues that most of the poaching in Garamba National Park in DRC was carried out by a combination of LRA, Janjaweed, and remnants of the SPLA.[79] Despite the headlines, the LRA is a minor player in elephant poaching and ivory trafficking. It has approximately 120 fighters, and numbers continue to dwindle. It does have a significant local effect on elephant populations in Garamba National Park, but it does not pose a major threat to elephant populations as a whole. It has been difficult to uncover exactly which groups have been responsible for poaching in Garamba, with park rangers reporting that they could be remnants of the Sudanese army, LRA, Janjaweed, or another of the many militias operating in the area.[80]

Indeed, one interviewee who worked in the Garamba National Park pointed out that the protected area has a 260km (162 miles) border with South Sudan, and so it is vulnerable to incursions by whichever armed group happens to be moving across the complex international borderlands; for them Garamba seemed surrounded on all sides by "ungoverned space."[81] Furthermore, another interviewee who worked in Garamba National Park emphasized that the key threat to wildlife was the illegal hunting of giraffes, not elephants, because the park contains the last remaining giraffe population in DRC. The interviewee stressed that different armed groups (including LRA) have to be flexible in order to survive in the area; this means they do not rely on ivory poaching to generate funding but become generalists who use a complex mix of poaching, gold trading, and looting.[82] Another high-level official in a donor organization put it like this: "They are just relatively poor people who are struggling to make a living and who will basically do something like that [poaching] on demand and know how to do it. These people are at one end of the chain. They are often people who have been associated with insurgent groups and armed groups in the past, so they're not necessarily different people at the poaching end of the spectrum."[83]

If the LRA are not heavily engaged in elephant poaching, why are they so often identified as the problem? As one US government official commented, the narrative that LRA is involved is very compelling and sexy. More importantly, invoking the LRA threat in the US context can bring in funding or garner congressional support.[84] But focusing on the involvement of armed groups diverts attention away from other groups involved in poaching and trafficking. For example, there are well-documented cases of armies and peacekeeping forces being engaged in poaching and trafficking. When the Congolese army was deployed to combat the LRA in Garamba National Park in the early 2000s, rates of elephant poaching actually increased because the army also took part in illegal hunting.[85] However, this story is not as headline-grabbing and may even tread on interstate or diplomatic sensitivities. By contrast, the story of LRA poaching ivory brings together a compelling mixture of a notorious militia group, the slaughter of elephants, and parks authorities battling to save wildlife.

In order to understand this rather better, we need to engage with wider debates about the LRA. Titeca and Costeur argue that the actors involved in combatting the LRA have framed the movement in ways that suit their own domestic ambitions and foreign policy aims.[86] For example, the Congolese government maintains the LRA is largely absent from DRC, a response shaped by the desire for the withdrawal of Ugandan troops from the country. Meanwhile, the US approach is informed by active and high-profile civil society campaigns, such as Invisible Children, Resolve, and the Enough Project, which aim to draw international attention. In essence, different actors framed the LRA, and the levels of threat it posed, in accordance with their own interests, either claiming it is a major threat, or very little threat, to stability and security in the border regions between Sudan, Uganda, and DRC. This thinking is echoed in debates about the possible role of the LRA in poaching as well.

Conservation NGOs, national governments, and international donors have sought to make the case that poaching is part of a conflict strategy for the LRA, in order to draw attention (and funding) toward conservation efforts and initiatives to combat the illegal wildlife trade. As one senior US government official put it: "There is clear evidence that wildlife products are traded for guns and ammunition by certain groups, certainly the LRA, they are hugely involved in that. There are other militia groups not just in CAR, but in northeastern DRC and South Sudan as well. So, they turn wildlife into guns and bullets with which they prosecute wars, hurt local people, destabilize governments, and they add to insecurity."[87]

Yet this picture does not map at all well onto the ways that the parks officials in Garamba explained the role of LRA in poaching in the area to me, nor does it tally with analyses by researchers who have worked in the area over a long period of time. It is interesting to explore how these arguments shape and change approaches toward conservation. They are central to the development of forms of conservation that are much more amenable to approaches and practices anchored in security, thereby producing a shift away from alternatives, for example, more community-oriented approaches in conservation.

Boko Haram is an important group to discuss here as it is also often referred to as a terrorist or militia group that targets wildlife. There are high-profile arguments that Boko Haram does engage in illegal wildlife trade, especially elephant poaching and ivory trafficking. For example, the Born Free Foundation USA/C4ADS report *Ivory's Curse* stated that elephants in Cameroon were poached by Boko Haram as the group moved across the international border.[88]

However, there are solid reasons to be very cautious about these claims that Boko Haram is trading ivory. Stories that mix together notorious armed groups and elephant poaching tend to appeal to external donors, international conservation NGOs, the media, and publics in wealthier countries. But such stories can distract from the need to focus on the actual source of funding for Boko Haram in Cameroon and Nigeria: cattle theft.[89] The idea of providing support and enhancing security for pastoralists in Nigeria and Cameroon does not excite the media and attract funders the way that stories about saving elephants while fighting terrorism do. Furthermore, a US government official with expertise in West African conservation pointed out that despite the ongoing popularity and influence of the narrative that poaching funds militia groups, it can be faulty in some places, including Boko Haram in Cameroon and Nigeria. The official suggested there was no hard evidence that Boko Haram is engaged elephant poaching, and since there are no elephants in that area anyway, it would be a faulty business plan. In fact, the poaching story was distracting from their actual activity, which is to capitalize on cattle rustling.[90]

The failure to deal appropriately with the changing nature of pastoralism in the region was also a feature of discussions with park authorities about Garamba National Park, who suggested that the source of insecurity was not militia groups and poaching; instead, "one of the security issues in this sub-region which is overlooked is pastoralism. As we see in Kenya and central Africa, transhumance does bring quite few issues that destabilise the region."[91] Failing to make this connection distracts from the genuine threats to wildlife in these areas. As one Washington-based conservationist pointed out "you don't have to look for terrorist groups or militants of various stripes to find the security problems . . . you don't need to have a cartoonish sort of bad guy to understand that we have a problem."[92]

Further, the notion that Boko Haram hunts elephants in Gabon to fund its activities is part of the justification for sending British Army trainers to work with park rangers, but Boko Haram is not active in Gabon. Following a visit from Professor Lee White, head of Gabon's National Parks Agency, a representative of one of the major donors for programs to tackle the illegal wildlife trade commented to me: "They know the criminals for wood, for wildlife, sometimes they are linked together, and they have other groups for fisheries. They have to deal with all the illegality that is happening. At the country level where we are working [terrorism] is not a security issue yet, . . . it is safe place, you can do tourism in Gabon. If it was terrorism we would not get clearances to go."[93]

Even donors can produce contradictory statements: for example, on the one hand, they fund projects in Gabon because they are convinced Boko Haram is operating as poachers yet, on the other hand, they are clear that terrorism is not an issue and that Gabon is stable and is safe for tourism.

In sum, the publicly available evidence of the illegal wildlife trade as a significant source of threat finance used by armed groups, militias, and even terrorist groups is not overwhelming nor is it completely convincing. Yet these stories have remarkable sticking power because they are compelling narratives, which intersect with other powerful agendas about global security and stability. They also neatly fit the objectives of national governments seeking international support for their own attempts to deal with armed threats within their borders. Furthermore, the arguments frame the illegal wildlife trade as a security threat, which produces a security-oriented response, thus deepening the integration of conservation with security. The framing of the illegal wildlife trade as a serious security threat is often taken at face value, as intuitively correct, and just seems like obvious common sense. Questioning these stories is often very unwelcome, but it is necessary because policy based on poorly evidenced and incorrect assumptions will be ineffective in the longer term. Of course, as I stated earlier, there could be evidence held by security agencies, which is not publicly available because of concerns about national security risks. If this is the case, it means policies that have the capacity to shape conservation in the longer term are not subject to wider public scrutiny, which produces wider challenges around transparency and accountability.

The high-profile claims that wildlife losses, poaching, and global insecurity are linked are very revealing. My argument in this chapter has not been to establish definitively whether illegal wildlife trade is indeed used to generate threat finance for armed groups and terrorist networks but rather to explore the debates about those links, raise questions, and examine how these debates are reshaping conservation. It is important to note that a wide range of organizations are communicating the same message in a very similar way, and that their arguments are based on a very narrow (publicly available) evidence base. It is also important to address what the underlying drivers of poaching and trafficking really are, and to question who is carrying out these activities, and to what extent, but a focus on specific armed groups may prevent an effective analysis. In making the link to global security, the underlying reasons for the appearance and activities of militia and rebel groups are left as a black box and not discussed. Further, linking illegal wildlife trade and armed groups taps in to contemporary anxieties about global security threats, the identification of legitimate targets for military action, and the seemingly endless US-led War on Terror. It is a significant shift from a war for biodiversity to a war by conservation, in which conservation can be rolled together with the US strategy for the War on Terror.[94]

There is much at stake. The idea that illegal wildlife trade constitutes a global security threat has reshaped conservation policy on the ground and has invited new kinds of actors into the wildlife conservation scene, actors who come with very different approaches and modes of thinking. The security-oriented approach to countering the illegal wildlife trade has produced important blind spots. Thinking of poachers as being in the same category as terrorists or armed groups, bent on creating instability, creates a powerful and convincing rationale for the use of more forceful and violent approaches. The relationships between people and wildlife are being restructured, so that it is deemed more and more acceptable to arm conservationists who feel threatened, to advocate for shoot-to-kill policies, and to enforce the separation of human and wildlife communities via violent means. These blind spots have important material outcomes. A strategy based on poorly evidenced assumptions that terrorism and trafficking are linked is likely to produce misplaced and ineffective responses to both terrorism and trafficking. If a policy is based on false assumptions, then interventions will be aimed at the wrong actors and activities, thereby overlooking which groups actually do engage in poaching and trafficking and why they might do so. Some have stressed the need to respond to an urgent situation, but the poaching crisis is now more than ten years in, and the response is still based on short-term, emergency thinking. If alternative, longer-term approaches had been funded ten years ago, they might now be yielding benefits, but we will never know.

5 SURVEILLANCE, INTELLIGENCE, AND CONSERVATION

The shift toward thinking of the illegal wildlife trade as a security issue has generated a growing interest in intelligence-led approaches as a means of tackling the illegal wildlife trade. In this chapter I examine and explain the growth of intelligence-led approaches that center on surveillance, monitoring, information gathering, and informant networks. These changes illustrate that the shift toward a security-oriented approach extends far beyond militarization, which is just one (albeit the most high-profile) expression of the growing integration of conservation and security. Rather like the militarization of conservation, intelligence-led approaches are not necessarily brand new, they have been used in the past; however, they are gaining greater momentum in part because of the ways that the wildlife trade is articulated as a serious crime and a global security threat in international networks. For example, in 2016, the International Fund for Animal Welfare (IFAW) sponsored a meeting with the American Geographical Society and the US Geospatial Intelligence Foundation (USGIF) in Washington, D.C., on the subject criminal nature: the global security implications of the illegal wildlife trade. It continued the focus on the findings of the 2013 report of the same name. The purpose was to "convene an innovative range of thought-leaders from conservation groups, U.S. Government, the policy community, and the private sector to translate the knowledge accumulated through research, policy, and practice into effective intelligence-driven action to conserve wildlife."[1] This statement is illustrative of a wider pattern. But what constitutes an intelligence-led approach to the illegal wildlife trade? What is its history in conservation? What challenges do conservationists face in data collection sharing? And how do intelligence-led approaches to the illegal wildlife trade intersect with those of investigating and prosecuting financial crimes?

WHAT ARE INTELLIGENCE-LED APPROACHES?

Intelligence-led approaches to tackling the illegal wildlife trade have gained importance for dealing with the trade's complexities. As one high-ranking ex-military intelligence officer, who worked for a high-profile conservation organization stated: "Developing strong networks that connect these vulnerable communities, wildlife service, and national and international agencies in order to efficiently and effectively build a comprehensive intelligence picture which can predictively drive a decision or action is one of the best methods we have to preventatively secure conservation landscapes and the people and wildlife within them."[2]

Intelligence-led approaches rely on collecting and collating information, strategically developing particular lines of inquiry and foci of surveillance, and working with complicated networks of informants. One intelligence specialist working in conservation, put it this way: "The gap between factual reporting and sensitive intelligence is very wide and [the two pieces] are not joining up. . . . In my long career in the military I saw exactly the same thing. I used to call it the hot tip brigade against the plodders who were just doing basic data collection and looking at trends and coming to conclusions."[3]

The faith in intelligence-led approaches has translated into the development of dedicated organizations to collect, collate, and share intelligence on poaching and trafficking. One good example is the Security Division of the KWS, which established an Intelligence Department in 2011.[4] This new department has "the responsibility of collecting, collating, and analyzing security related information from which it generates intelligence concerning the security of wild flora and fauna, security of tourists and KWS resources. The department undertakes surveillance and monitoring of bandits and gangs around wildlife-protected areas."[5]

In broad terms, the debates about intelligence-led approaches can be divided in to two overlapping, but sometimes contradictory, sets of issues. First, intelligence networks and use of surveillance are central to counterinsurgency techniques[6] and can be important elements in developing and maintaining more militarized and forceful approaches to conservation. The US military refers to this as "irregular warfare," a term that now features in conservation discussions, especially in how to counter illegal wildlife trade. Part of the explanation lies in the ways that the integration of security and conservation facilitated the entry of new kinds of actors into conservation, especially US war veterans. One US veteran who trains rangers remarked, "The model that we use when it comes to interacting with local rangers, with the community, with the law enforcement, is what we would call an irregular warfare model. It's 4th generational warfare."[7]

Counterinsurgency techniques are designed to tackle groups that seek to overthrow a defined enemy and extend territorial control. Kilcullen argues that, classically, the goal of insurgencies is to overthrow a state, but many insurrections in contexts of protracted conflict do not seek to take over a functioning state apparatus; instead, they aim to capture control over remnants of it or gain control of key territories or resources. Counterinsurgency strategies adapt to respond to the shifting tactics of insurgencies.[8] Gaining control of the state or its remnants is not the primary aim of poachers, since they are not necessarily interested in overthrowing the state or extending territorial control. Instead, their aim is to extract a valuable resource and trade it for economic gain. So in this regard, using Kilcullen's understanding of insurgent groups, organized groups of commercial poachers could be regarded as akin to insurgents because they seek to gain access to and control over lucrative resources by use of force;[9] however, not all poachers are necessarily akin to insurgency groups operating in conflict zones.[10] Nevertheless, counterinsurgency strategies have been deployed in a range of conservation contexts, and especially in initiatives focused on the illegal wildlife trade.

Second, intelligence-led approaches are sometimes presented as an alternative to the use of force and militarization because they rely on informants, policing tactics, data collection, and crime prevention. Intelligence-led responses to the illegal wildlife trade are seen as being integrated with upholding the law, which includes the use of the police, judiciary, and international police agencies. They fit well with the ways that the illegal wildlife trade is increasingly referred to as "wildlife crime" or as carried out by transnational organized crime networks. One security specialist, who provides military and intelligence training for private reserves, described how intelligence gathering was central to the initial stages of developing a training package for conservation staff: "I will take a security element with me to a reserve. It's usually a plus one, so it's not very heavy. We just go and observe. Prior to me going to even make that observation, I will build what's called an OSINT. That's an open-source intelligence summary. I pore through as many things as I can, get as much information as I can about stakeholders, key players, specific individuals, criminal behavior, even going as far as I can to contact with local law enforcement."[11]

Similarly, an activist involved in undercover operations to tackle the illegal wildlife trade stated that their approach was a unique model, "covering the whole chain of wildlife crime: it investigates, which means investigators go undercover so we get information from inside the networks we are fighting against. This helps understand more how they operate and strategize. We carry out arrest operations, which means we can go along with the arresting authority to increase the success of the arrest, we can also carry [out] intelligence activities after the arrest, . . . analyzing seized phones, documents, house searches, etc. We can also assist the authorities during interrogation of suspects so we can assist with questions that possibly the local authorities aren't aware of since they are not experts, among other advantages."[12]

However, intelligence-led approaches and militarization are not mutually exclusive categories; there are overlaps that need further exploration. Intelligence-led approaches do have some similarities with militarized approaches, but it is too simplistic to think of them as the same thing.[13] Policing can mirror military approaches in instances where the police are heavily armed, able to use deadly force (whether authorized or not), and are kitted out with body armor and body cams. In these circumstances, it can appear that there is little difference between civilian and military organizations involved in enforcing conservation laws.

In many countries, the police are a significant and forceful arm of the state, with functions that include enforcing government control over resistant populations. Further, in some places the military are also responsible for policing. For many communities around the world, their experience of interactions with the state is one of oppression, violence, and marginalization. There have also been more general shifts in policing across the world toward intelligence-led approaches and use of technologies, surveillance, and informant networks. Therefore, in calling for the use of policing to tackle wildlife crime, it is important not to assume a universal model whereby a policing approach is carried out by a politically neutral force, upholding a just law against criminal transgressors. The framing of illegal wildlife trade as a serious and organized crime can encourage and prompt responses that are anchored in approaches from policing, in much the same way that referring to it as a security threat leads to security-style responses. In short, the language used to describe illegal wildlife trade really matters in terms of shaping and driving policies on the ground.

For example, in Laikipia, Kenya Police Reserves (KPR) is separate from the Kenya Police Service, and it is an auxiliary force formed of unpaid volunteers operating within their own areas. KPRs are provided with arms by the state to support policing where police presence is low.[14] Furthermore, Northern Rangelands Trust (NRT) developed three elite rapid response units, named 9–1A, 9–1B, and 9–2, that have received enhanced training from the private military company 51 Degrees, which is associated with the son of the NRT director, Batian Craig.[15] They have a broader remit to tackle cattle rustling, banditry, and inter-communal violence in the area, as well as poaching.[16] The elite units are drawn from all three main ethnic groups in the area and are often pointed to as an example of best practice.[17] The NRT has attracted donor support for the development of a complex security apparatus as part of the wider war on poaching. An example, in 2013, was US support for trials of drone use by the US company Airware. The deployment of high-tech surveillance in the area was suspended in 2014 when the Kenyan government, prompted by concerns about state security, ended all rights for private sector to operate drones.[18]

However, Mkutu and Wandera argue, the KPRs were not effective at providing security.[19] As a result, the Kenyan government contracted the British Army, which has one of its largest training bases in the area, to provide professional training. Further, critics have argued that such initiatives do not address the ways in which poaching in Kenya is linked into much broader sociopolitical dynamics, including cattle rustling, road banditry, and intercommunal conflict, with individuals and weapons rotating between different groups involved in these activities.[20] The provision training by the British Army and by private security firms, as well as the allocation of KPRs to the Laikipia area, indicate ways that the resources of states can be allocated to protect private interests (such as the Conservancies in Laikipia) and further extend the objectives of state security and stabilization for large areas of Kenya.[21] Counterinsurgency strategies can have wider uses when deployed in conservation—they can become enrolled in the broader objectives of states and the private sector by producing security for particular (powerful sets of) groups and interests.

Proponents of security-oriented approaches argue they are necessary because poachers are becoming more heavily armed. This raises the question: if poachers are becoming more heavily armed, where are the weapons coming from? In the case of Kenya, Maguire suggests these additional arms are procured via thefts from British Army training camps; while there is evidence that some poachers are using more powerful weapons, KWS reports indicate that between 2012 and 2014 there was an increasing trend toward the use of cheap and low-tech methods, including traps, arrows, and poison to avoid detection.[22] More broadly, Mbaria and Ogada suggest that the model of the NRT facilitates the continued inequalities in land distribution in Kenya. Some in the communities find the public statements made by NRT patronizing, especially the suggestion that funds are needed to engage in community education about the value of wildlife and conservation, as if they did not already hold any valuable knowledge about wildlife.[23]

Further, the enhanced levels of information gathering, monitoring, and surveillance capacities of nonstate actors like conservation NGOs or private conservation initiatives are worth greater reflection. One interviewee who runs a private sector intelligence company working with conservation projects commented that "the whole world of intelligence is shrouded in, I think, misperception . . . security and conservation are becoming integrated and this may not be regarded positively by some organizations. I very much accept that the words security and intelligence have bad connotations."[24]

Intelligence-led approaches are gaining wider acceptance and are often presented as distinct from, or in place of and preferable to, militarization. This is in part explained by the framing of the illegal wildlife trade as an urgent crisis, a serious crime, and a global security threat. This way of thinking lifts intelligence-led approaches out of the realm of normal politics, as securitization theorists might put it. It renders conservation part of the politics of exception: an urgent situation requires a security response, which is placed beyond oversight and scrutiny because of concerns about national security and the sensitivity or confidentiality of material obtained via intelligence gathering.

THE HISTORICAL DEVELOPMENT OF INTELLIGENCE-LED APPROACHES

The intelligence-led approach builds on important historical antecedents, and so the current phase is not necessarily a brand-new dynamic. Rather, conservation has a long history of being integrated with gathering intelligence for states and nonstate agencies alike. These past experiences reveal important lessons for current practices. One well-documented example of an attempt to use intelligence-led approaches to tackle poaching is Operation Lock.[25]

In 1987 Prince Bernhard of the Netherlands approached John Hanks, the then-director of Africa Programs for WWF-International, to ask whether there was an organization that could undertake covert operations to tackle poaching and trafficking of rhino horn. Prince Bernhard was keen to ensure that the initiative was kept off the official books.[26] Hanks approached a new private security firm, KAS Enterprises, led by Sir David Stirling, founder of the British Special Air Services (SAS). After several discussions it was agreed that Colonel Ian Crooke, a former SAS officer, would head up a small team of former armed forces and intelligence operatives under the name Operation Lock. Once in place the team developed a strong network of informants and collaborators across Southern Africa, provided training to conservation authorities, and began to infiltrate illegal trading networks in the region. The undercover methods it employed meant that in order to infiltrate the illegal networks, the team had to pose as smugglers of ivory and rhino horn, and they obtained 178 rhino horns in order to do so.[27]

Internal documents reveal that KAS Enterprises believed it could become financially self-sustaining by selling illegal wildlife products, yet if it has done so, this would have made it one of the largest smuggling rings in Africa.[28] At the same time Crooke also collaborated with South Africa's superspy, Craig Williamson, who was running a covert operation against the African National Congress (ANC) for the apartheid regime. There were concerns among some government officials in the neighbouring countries that Operation Lock was collecting wider intelligence to assist the apartheid regime in its campaign to destabilize the Front-Line States.[29]

Until recently, John Hanks maintained that Operation Lock was conducted without the knowledge of WWF-International, despite ample evidence to the contrary; the suspicion remains that Operation Lock was used by the apartheid regime to develop operations against the African National Congress.[30] In 2015 John Hanks published a book, *Operation Lock and the War on Rhino*

Poaching, to give his perspective on the whole affair and the fallout that followed. In the book Hanks makes it clear that while the activities of Operation Lock fell outside WWF-International's official remit, staff there were fully aware of the operation. Hanks also details his feeling that he had been set up as a fall guy for WWF-International in the scandal that ensued.[31] Operation Lock ended in 1990, after an exposé by a Reuters journalist. A year later, many of the responsibilities, activities, and networks from Operation Lock were taken forward by South Africa's newly formed Endangered Species Protection Unit.[32]

Operation Lock is just one of several examples which predate the current interest in intelligence-led approaches to tackling the illegal wildlife trade. Yet it is indicative how well-intentioned plans to save wildlife can be integrated with wider objectives of states and private companies.

Several conservation NGOs continue to hire advisors trained in military or police intelligence, or partner with private intelligence companies staffed by former operatives from intelligence services, such as Bureau of State Security in apartheid South Africa and HaMossad leModi'in uleTafkidim Meyuḥadim (MOSSAD) intelligence agency of Israel.[33] These include WWF, IFAW, Wildlife Conservation Society, Zoological Society of London, Biglife Foundation, Born Free Foundation, and Space for Giants. As a result, conservation NGOs are developing intelligence-gathering capabilities in partnership with private sector military and security companies as well as with governments across the world. There are long-term risks and dangers in this for people, wildlife, and conservation NGOs alike. For NGOs, there are reputational risks of partnering with the security sector, as Operation Lock indicates. They can be engulfed in scandals, which then affect their ability to raise funds and operate in particular locations. But what challenges face conservationists using these intelligence-led approaches?

DATA GATHERING AND SHARING

Gathering and sharing data on the illegal wildlife trade, the geographical patterns of the trade, the individuals and groups involved, the transport routes, and entry and exit ports is vitally important to intelligence-led approaches. UNEP, for example, states that monitoring the legal trade and curbing the illegal trade in wildlife requires good information exchange and cooperation, involving importing, exporting, and transit countries. Therefore, UNEP recommends that mechanisms need to be enhanced to facilitate rapid exchanges of intelligence among enforcement agencies.[34] When he was appointed as chief of enforcement for CITES, John Sellar noted that the data provided by CITES member states was often "so haphazard and incomplete that it became almost meaningless."[35] When data from CITES, INTERPOL and the World Customs Organization were compared, there was very little overlap between datasets, which indicated that each organization had its own distinct sources of information.[36] This lack of information exchange remains a problem, and perhaps a growing one, for tackling the illegal wildlife trade. One of the key challenges is guaranteeing anonymity for anyone who acts as an informant, which is not easy to do. Intelligence-led approaches rely on strong and trusted informant networks, and they only work if the informants can be confident that their identities will not be revealed.[37]

National authorities, NGOs, and intergovernmental organizations, or a combination of them, all engage in monitoring the legal and illegal wildlife trades. Among these are relevant government ministries, the national authorities responsible for CITES, EU-TWIX database, national police forces, INTERPOL, and TRAFFIC International. For example, the World Conservation Monitoring Centre (WCMC)–CITES database reports all records of import, export, and re-export of CITES-listed species as reported by parties.[38] Because of the range of organizations involved in monitoring and data collection, there are several barriers that need to be overcome. These include different reporting and operating systems, regulations about confidentiality, as well as territorial turf wars over which organization holds what information or lack of knowledge about what information is held and by whom.

Despite these challenges, some existing mechanisms encourage greater degrees of information exchange. One of the most important sources of data for the EU is the EU-TWIX database, which was established in 2005.[39] It is funded by the European Commission and is currently managed by TRAFFIC (Brussels office), which runs it for use by EU enforcement officials. It uses CITES Alerts for rapid exchange of information on seizures in member states, and access to the data is normally restricted to enforcement officials. One example of the benefits of effective information exchange is the 2010 arrest by customs officials in Guyana of a Dutch national who was smuggling hummingbirds. The customs authorities in France had information, which they shared via EU-TWIX, allowing Dutch authorities to trace a wider network engaged in smuggling the birds.[40]

A second example is provided by EUROPOL, which collects and collates information to support the law enforcement initiatives of member states. EUROPOL recognizes that the transborder nature of environmental crime means there is a need for a pan-European response. Indeed, environmental crime is a priority of the European Multidisciplinary Platform against Criminal Threats (EMPACT) for its policy cycle of 2018–21, with a particular focus on the illegal wildlife trade and on trafficking of waste.[41] EUROPOL provides the permanent secretariat for the Environmental Crime Network (EnviCrimeNet), which is an informal network connecting police officers and other crime fighters in the field.[42] In 2015, EnviCrimeNet and EUROPOL finalized a year-long intelligence project on environmental crime, using data from fifty jurisdictions. It concluded that the main issue for tackling environmental crime is that much of it goes undetected due to the reticence or inefficiency of law enforcement agencies in dealing with this problem.[43]

Unlike drug trafficking and other forms of crime, which over thirty or more years has developed a cooperative (though not perfect) infrastructure for data collection and sharing among law enforcement communities, there is no such framework for environmental crime.[44] A senior international law enforcement representative commented: "You have a plethora of agencies involved, police services, customs services, management authorities involved. When we talk about environmental crime you have various forms of environmental crime: wildlife crime, waste trafficking illegal fisheries, illegal logging. And there are different authorities for each. The cooperation and information exchange is less than optimal. Then you need to assume they are generating proper data."[45]

The enthusiasm among a wide range of NGOs for data collection on the illegal wildlife trade is an indicator of their commitment, but it does produce problems for data sharing and collation. A senior international law enforcement officer captured this beautifully: "I can only compare it with a football field, where everyone has their own ball and is playing with it. It is only by accident that I get to know things are happening. Nothing is structured or really coming together."[46]

This comment demonstrates that the proliferation of databases and organizations collecting important intelligence on the illegal wildlife trade has not been coordinated, leading to duplication of efforts, lack of information sharing, or even turf wars over which organization has the right to collect and store what information. These same sentiments were echoed by the representative of an intelligence-gathering network of conservationists, working in East Africa, who commented on the challenge of big egos in conservation: "While we call syndicates 'organized criminal networks' I would say the conservationist or conservation NGOs are 'disorganized networks.' We are losing the battle just on this basis. If we want to fight the top masterminds of these crimes we need to begin sharing information and success stories. We need to stop the backstabbing and non-constructive critics [*sic*], as well as also being able to listen and accept the constructive critics [*sic*]."[47]

Data on seizures of illegally traded wildlife products are very important for understanding trafficking routes, the kinds of products being traded, and the effectiveness of strategies to detect and intercept illegal consignments. There are several sources of information on seizure data. First, there is the CITES trade database which holds information on legal trade in wildlife, as well as data on seized and confiscated goods; second, there is the EU Trade in Wildlife Information Exchange (EU-TWIX) database for all EU seizures; and third, the Law Enforcement Management Information System (LEMIS) for US seizures.[48] News about seizures do

Seized lion head displayed by Australian Border Force at the IUCN World Parks Congress, Sydney, Australia, 2014. Photo credit: Rosaleen Duffy.

make headlines and get widely reported, especially when they are large-scale seizures of charismatic animals, like elephants or highly endangered species such as pangolins or tigers. This information is also compiled by a range of NGOs in the form of reports on seizures and the profiles of the illegal wildlife trade. For example, TRAFFIC-International has compiled a thorough list of reports of seizures from 1997 to 2019.[49] Large-scale seizures of high-profile species hit the headlines as good news stories about the successes of efforts to tackle illegal wildlife trade. The Environmental Investigation Agency (EIA), for example, estimated in 2019 that "Vietnam has overtaken mainland China as the world's leading destination for illegal ivory, having made 38 large-scale ivory seizures during the reporting period. In March 2019, over nine tonnes [ten tons] of ivory was seized in Hai Phong, Vietnam, in the world's largest ivory seizure to date. The following countries made the next highest number of large-scale ivory seizures after Vietnam: Hong Kong SAR (17), Kenya (14), mainland China (12) and Thailand (11). The large-scale ivory seizures mapped here correspond to a total of approximately 293 tonnes [323 tons] of ivory, equivalent to an estimated 43,840 dead elephants."[50]

There are problems with relying on seizure data (when national customs authorities intercept and report shipments of illegally traded wildlife) as a measure or indicator of levels of the illegal wildlife trade and success or failure of law enforcement responses to it. As Tittensor et al. argue, seizure data represent an absolute lowest possible level of illegal trade activity since so much wildlife that is illegally traded goes undetected.[51] EIA suggests that as many as one million pangolins have been illegally traded within Asia in the past ten to fifteen years, but it recognizes that the figures on seizures are an underestimate and could represent as little as 10 percent of actual illegal trade, such that the total number of pangolins traded could be nearer one and a half million during the period.[52] Seizure data can also reflect the robustness of national law enforcement. For example, for several years CITES data showed Austria as the country with the highest level of illegal wildlife trade in the world, but it was merely a reflection of its systematic and higher-quality reporting of seizure data.[53] Seizure data is an important snapshot of the trade, giving information about animals that have already been captured or killed and then trafficked. Higher levels of seizures can reflect one of two things: that illegal trading in a particular animal or animal product is actually increasing, or that enforcement efforts to intercept goods have been more successful. Further, seizure data does not reveal the situation on the ground in places where wildlife is extracted or is consumed.

As part of intelligence gathering and information exchange, conservation agencies and NGOs have also made use of mobile technologies and opportunities for surveillance of individual users. For example, in Kenya mobile phone data was used to develop intelligence on the wider network of people who send, receive,

Demand reduction campaign by WildAid and Change, Vietnam. Photo credit: Anh Vu.

store, and use money linked to wildlife crime. This can be especially useful in developing a bigger picture of international networks of traffickers who use mobile phones to communicate with one another. In the case against Feisal Mohammed Ali in Kenya, authorities used mobile phone data (including call logs) to link him to a much bigger smuggling ring.[54] Even in primarily cash-based economies in rural areas, use of mobile phone payment systems can also be an important source of information on poaching and trafficking networks.

Mobile money has been used to facilitate transactions related to wildlife crime at the lower end of the chain. This has encouraged conservation NGOs to partner with mobile phone networks, law enforcement agencies, and private sector companies engaged in intelligence gathering to develop ways of using mobile phone networks to detect wildlife crime. For example, the mobile network Safaricom in Kenya has developed sophisticated in-house intelligence functions to support their mobile-money businesses, using call log meta-data and screening software to monitor for high-risk users. Safaricom has joined the information-sharing taskforce established by Kenya Wildlife Services, which allows Safaricom to refine its monitoring of users in line with KWS priorities.[55] The ways that mobile phone operators and wildlife agencies are working together is indicative of the security turn in conservation. It fits with the wider arguments that the there is a need to respond to an urgent crisis, which requires drawing in a different range of actors than the norm (such as mobile phone network operators).

Intelligence gathering to tackle the illegal wildlife trade has generated new kinds of cooperation agreements between enforcement agencies and wildlife conservation NGOs. For instance, IFAW and EIA cooperated with INTERPOL on Project Wisdom and Project Worthy, which include training for enforcement agencies dealing with wildlife trafficking.[56] EUROPOL signed a memorandum of understanding (MoU) with TRAFFIC International in 2016 which was intended "to facilitate the exchange of information and support, as well as to improve coordination between the two organizations to fight environmental crime, particularly the illegal trafficking of endangered animal and plant species."[57] Similarly, on 13 March 2017, EUROPOL director, Rob Wainwright, and Wildlife Justice Commission (WJC) executive director Olivia Swaak-Goldman signed an MoU agreeing to share technical expertise and best practices and to raise awareness of environmental crimes, including trafficking in endangered animal and plant species.[58] WJC has an Intelligence Unit, which collects, collates, and analyzes information to develop tactical, operational, and strategic intelligence assessments for law enforcement at a national level—and also in support of efforts by international agencies.[59]

This raises the question of how to determine the appropriate level of access to data for NGOs, as well as sharing sensitive data on criminal activity or security concerns. WJC and EUROPOL have had to grapple with this issue in their partnership agreement. WJC clearly states that its role is "not intended to replace or circumvent domestic processes."[60] EUROPOL is clear that collaboration with

WJC "can build up capacity in the fight against the trafficking of endangered animal and plant species."[61] However, because there are strict protocols surrounding sharing of information at EUROPOL, the agency does not provide reciprocal information to the NGOs. As a EUROPOL official remarked "We appreciate NGOs, . . . they can provide very relevant strategic information. But what they can get from our side is, in terms of data, basically nothing."[62] The collaborations with law enforcement agencies have additional benefits for the NGOs. Conservation NGOs may seek such collaborations and MoUs in order to enhance their standing and legitimacy with key donors. The ability to demonstrate a close working relationship with, for example, EUROPOL is important when NGOs seek funding from the European Union for their activities and campaigns.[63] Furthermore, conservation organizations can see themselves as offering oversight to reduce the chances of corruption in investigating and prosecuting wildlife crimes. As one conservationist actively involved in criminal investigations stated, "During all of these steps [NGO name] is basically serving as a 'third eye' also to avoid any attempt of corruption throughout the process. Offer of bribes takes place as soon as the suspect/s are arrested, from the local police and all the way to the magistrate, even to jail guards. If you have someone 'looking over your shoulder' it will be a bit harder to commit corruption."[64]

One further example of the confluence of those trained in the intelligence services with conservation is Ofir Drori, who set up the Eco Activists for Governance and Law Enforcement (EAGLE) Network. He has briefed MEPs at the European Parliament about the illegal wildlife trade. One MEP commented: "He lives in Cameroon, and he has amazing networks of people, he was trained by Mossad. He uses local informants in villages and he is getting some really good results, taking Generals and MPs to court."[65] The EAGLE Network has developed a model of working with governments on investigations, arrests, legal follow-up, and media attention. It also focuses on detecting corruption in government services and identifying weak law enforcement and judicial ineffectiveness as key to sustaining the illegal wildlife trade. The

EAGLE Network started in Cameroon and has extended to a further seven countries (Congo, Gabon, Guinea, Togo, Benin, Senegal, and Ivory Coast).[66] The EAGLE Network states that it uses undercover methods, and agents (whose identities remain anonymous) to infiltrate criminal networks and can set up sting operations. Further, in situations where it has planned and arranged arrests, it supervises the arresting forces (state agencies) in order to ensure that corruption does not jeopardize prosecutions.[67] It is an interesting example of an NGO that has authority over state agencies in certain situations. It differs significantly from examples of collaborations between wildlife conservation NGOs and larger international law enforcement networks, such as EUROPOL and INTERPOL. It is also indicative of the power dynamics and patterns of accountability in some intelligence-led approaches to tackling the illegal wildlife trade. There are very few arenas where governments will allow nonstate actors to exercise this kind of authority within their borders, especially in matters that could affect national security. That is the power of wildlife conservation: it is put above "normal politics" and can facilitate collaborations or even outsourcing of intelligence gathering to nonstate entities, especially in cases where state capacity is low.

The development of intelligence-led approaches has produced important and influential partnerships between state agencies and conservation NGOs. One of the most high-profile examples is Operation tenBoma, established in 2015 as a joint initiative of the KWS and IFAW. It is an example of how intelligence-led approaches can be inspired and informed by military intelligence practices. Operation tenBoma was originally the idea of Lieutenant Colonel Faye Cuevas, IFAW senior vice president, and James Isiche, East Africa regional director for IFAW.[68] Cuevas has a military background, having served in military intelligence for the US Air Force, and has been involved in missions in the Middle East, and Africa. As part of the Special Operations Command-Africa, Cuevas led intelligence, Surveillance, and Reconnaissance support for more than one hundred special operations missions in West Africa, Central Africa, and the Horn of Africa.[69] She brought this background and military thinking to conservation, and the result was tenBoma.

The idea behind Operation tenBoma was to develop a network of local and global communities to gather and share intelligence in order to prevent poaching activity. IFAW states that the techniques used by Cuevas in military intelligence operations, including "more robust technical components and more effective communications, information processing and analytical methodologies" are the "same proven methodologies" which serve as the foundation of tenBoma.[70] It is a good example of how approaching conservation as a security issue prompts and facilitates a security-type response to it. It draws on military intelligence techniques to tackle poaching and trafficking, which involves greater integration and two-way transfer of knowledge, techniques, and approaches between security and conservation. Indeed, one ex-military trainer working with rangers in private reserve indicated how he thought that military-style intelligence training led to immediate benefits: "The human behavior pattern recognition and the human terrain mapping that —— teaches, the very first time that he taught that in Africa, the rangers went on a lunch break the first day. On their way to lunch, down the road outside of the reserve, they caught eight individuals. They saw some of the behavior patterns that they had just learned in the class, stopped them before they could even get over the fence, before they could get to their weapons, and just had a conversation. They dictated, based on what they learned, that these individuals are suspicious and they're showing behavior patterns that somebody who has something to hide, basically, would be showing. Sure enough, they were going to go out and poach. Boom, just like that."[71]

It is clear that intelligence-led approaches can be effective in certain circumstances, and that they can assist conservation authorities in detecting or preventing illegal activities and in capturing and prosecuting poachers and traffickers. More broadly, the transfer of military intelligence-gathering techniques to conservation practices is an interesting development, facilitated by the redefinition of illegal wildlife trade as a serious form of wildlife crime.

INTERNATIONAL ORGANIZATIONS AND POLICING APPROACHES

The idea that the illegal wildlife trade can be tackled via intelligence-led approaches fits well with arguments that there is a need for better international cooperation in policing and law enforcement. INTERPOL has been involved in several high-profile initiatives on illegal wildlife trade, in collaboration with conservation NGOs and with other international agencies from the UN system. INTERPOL has 190 member countries and each member has a national central bureau. However, INTERPOL does not have policing powers of its own and instead acts as a coordinating agency. Its Environmental Crime Program acts as a coordinating hub for responses to the illegal wildlife trade. It also developed the Eco-message system to facilitate information exchange between members and to analyze trends in environmental crime and better target enforcement.[72] In cooperation with EUROPOL, INTERPOL supports enforcement through its Environmental Crime Committee and Environmental Compliance and Enforcement Committee. Conservation NGOs have proved to be important collaborators in INTERPOL's monitoring and enforcement efforts.

EUROPOL started to work on illegal wildlife trade much more recently, first in a coordinating role, which has now developed into a much more significant involvement. A good example is Operation Cobra III in May 2015. It was the largest ever coordinated international law enforcement operation to target the illegal trade in endangered species and included law enforcement teams and agencies from sixty-two countries in Europe, Africa, Asia, and the United States. EUROPOL facilitated this program by coordinating connections and information exchanges among the police, customs, forestry, and other law enforcement authorities from twenty-five participating EU member states. The operation was organized by the regional organizations Association of Southeast Asian Nations Wildlife Enforcement Network (ASEAN-WEN) and the Lusaka Agreement Task Force (LATF).[73]

This shift in EUROPOL followed interest and pressure from networks of practitioners, law enforcement, NGOs, and MEPs.

Until 2016–17 the illegal wildlife trade had not been a major focus for EUROPOL. In 2013 it published a *Threat Assessment on Environmental Crime in the EU,* which noted the emerging threat posed by trafficking in endangered species in terms of impact, high-value, modus operandi, and dimension in the EU as well as worldwide.[74] EUROPOL drafted the Serious Organized Crime Threat Assessment (EU SOCTA), which identified the priority threats to European security. The EU SOCTA determines how funding is allocated in each four-year policy cycle.[75] In 2013 there was not enough information in the databases to define environmental crime as a serious threat and a priority. However, "on the political level in Brussels, the Council of Ministers, they still decided it was a necessary an area to focus on, so environmental crime was included as a priority for EUROPOL because of some political pressure."[76] A MEP involved in persuading the EU to take the issue more seriously and allocated more funding stated, "We had a real push from here with all the MEPs writing to the Environment Ministers and Justice Ministers from across the 28 members [to ensure that] when it came to setting the budgets for SOCTA to make sure there was money in the budget for that."[77] These coordinated efforts produced the desired effect, and the illegal wildlife trade was identified as a priority area for the 2018–21 policy cycle. Despite this, a senior law enforcement official stated in 2017 that the EUROPOL knowledge base was still in its infancy and remained quite limited.[78] In 2020 the EU launched a review and called for responses to the *Action Plan against Wildlife Trafficking,* in readiness for developing the next phase. In discussions with MEPs at a knowledge exchange workshop that I chaired at the European Parliament in January 2020,[79] it became clear that illegal wildlife trade would be an even bigger priority in the next policy cycle. In sum, intelligence gathering is a central strategy for law enforcement agencies seeking to tackle the illegal wildlife trade. Despite the limitations and challenges they face, such approaches allow for cooperation across different law enforcement agencies, including information sharing and working together on specific operations to intercept shipments or detect and prosecute traffickers.

ILLEGAL WILDLIFE TRADE AS A FINANCIAL CRIME

The idea that the illegal wildlife trade generates threat finance has prompted some organizations to call for using existing instruments aimed at financial crimes, including anti-money laundering techniques. Historically, those concerned with anti-money laundering have not shown a great deal of interest in wildlife trafficking.[80] However, this has changed as a result of the growth of the idea that the trade is a serious crime and a security threat, and there is much greater interest in adopting some of the approaches used to tackle financial crimes.[81] This sits well with the wider integration of conservation with security concerns too. Indeed, at the 2018 London conference there was a clear emphasis on using financial instruments to tackle illegal wildlife trade. The phrase "follow the money" was used repeatedly to emphasize how the existing regulations concerning finance and money laundering could be used to detect and prosecute those engaged in illegal wildlife trade. Further, at the conference the UK government formally launched the United for Wildlife Financial Taskforce, pledging £3.5 million in assistance, and the Duke of Cambridge, among others, referred to the ways that Al Capone was prosecuted for tax evasion, not murder, and thus *Al Capone tactics* were promoted as a means of tackling illegal wildlife trade.[82] It is a fitting label because, as noted by Sellar, the large profits from illegal wildlife trade have led to the involvement of organized crime networks.[83] Despite the fact that anti-money laundering and asset recovery has been successful in tackling some criminal networks, to date hardly any of these instruments had been used against wildlife criminals.[84] For example, the Wildlife and Forest Crime Analytical Toolkit compiled by the ICCWC demonstrated that by 2012 the attempts to track money laundering by freezing and confiscating the proceeds of wildlife and forest crime had been undertaken only within one country, and not internationally.[85]

As Wittig suggests, CITES is not well equipped to deal with organized crime networks because CITES authority ends precisely where the involvement of organized crime begins.[86] And there is very little common agreement on what constitutes a financial crime or an environmental crime.[87] In the arena of wildlife trafficking, laundering itself has a wider definition because legal companies and forms of trading can be used to transform the profits derived from illegally obtained and traded wildlife into a legal and legitimate business. Further, individuals or groups within legitimate businesses or official organizations dealing with wildlife can be involved in the illegal trade and use their positions to launder wildlife, wildlife products, or profits from trafficking. Dan van Uhm calls these "green collar crimes".[88] Therefore, the sole focus should not be on those defined as part of the criminal underworld, but rather on the green collar perpetrators in captive breeding centers, zoos, and commercial trading companies, who are knowledgeable and skilled at exploiting legal loopholes and evading capture.[89]

In order to understand how anti-money laundering or financial crime approaches may assist in tackling the illegal wildlife trade it is useful to set out the main international instruments that can be used. One of the main intergovernmental organizations involved in addressing money laundering and terrorist financing is the Financial Action Task Force (FATF), established in 1989. The FATF is intended to set standards and promote implementation of regulations.[90] Despite the efforts of the FATF to develop standardized definitions, money laundering can mean anything from using funds gained through crime to the complex financial transactions used to hide the criminal origins of finance.[91] Further, there is little knowledge of the regulations and how they apply to wildlife trafficking. The Eastern and Southern Africa Anti Money Laundering Group (ESAAMLG) of the FATF reported in 2016 that there was little knowledge of the role and purpose of FATF's Financial Intelligence Unit (FIU), or of the support it could provide in tackling money-laundering activities as a means of combating wildlife crime.[92] Indeed, in investigations into other types of crimes, such as trafficking humans, drugs, and weapons, tracking the financial element is regarded as an important method for identifying individual criminals, as well as for gathering intelligence on their wider network. However, it is not employed in a systematic way for investigating wildlife crime.[93] Haenlein and Keatinge point out that "in the majority of wildlife crime investigations, financial leads go unfollowed, financers and supporters unidentified and assets unfrozen and un-confiscated. In large part, this is because of a delay in recognising wildlife crime as financial crime. . . . But it is also due to a lack of understanding of exactly how money linked to wildlife crime moves and what signs to look for."[94]

Approaching the illegal wildlife trade via existing mechanisms to tackle money laundering and financial crime has been advocated by international organizations. For example, in 2013 the United Nations very clearly drew the link between the increasing sophistication of wildlife trafficking networks and other serious offenses, including theft, fraud, corruption, drug and human trafficking, counterfeiting, firearms smuggling, and money laundering.[95] Similarly, in April 2015, the UN Congress on Crime Prevention and Criminal Justice (held in Doha, Qatar) issued the formal Doha Declaration, which recognized wildlife trafficking as a serious form of transnational organized crime. Further, it recommended strengthening legislation, international cooperation, capacity-building, criminal justice responses, and law enforcement efforts aimed at, inter alia, dealing with transnational organized crime, corruption, and money-laundering linked to such crimes.[96] In the same year the UN General Assembly adopted its first-ever resolution on wildlife trafficking, which also stated that member states should review and amend legislation on wildlife trafficking to ensure that it is treated as a serious crime, one that can be treated in the same way as money laundering under the United Nations Convention on Transnational Organized Crime (UNTOC).[97] The idea of using such tactics as "follow the money" and financial regulations to tackle illegal wildlife trade are gaining ground; they have been boosted by the redefinition of the trade as serious or organized crime. It facilitates the involvement of a wider range of actors to tackle wildlife crime—from financial regulators to international law enforcement and intelligence specialists. This approach is now increasingly supported by a range of international organizations, conservation NGOs, and private sector actors.

Intelligence-led approaches are part of the wider integration of conservation and security as a result of framing the illegal wildlife trade as a security threat. The development of the notion of wildlife crime as a serious crime means that it now has an enhanced status in international policy-making circles, and this designation also allows for and facilitates high-level cooperation and information exchange which was not possible before it was defined as a serious crime. Such coordination and information exchange are vital for tackling the complex chains of different actors involved in the illegal wildlife trade. Intelligence-led approaches should not be thought of as a singular kind of practice because they can overlap with forms of militarization of conservation and because the use of informant networks and counterinsurgency techniques are also features of some forceful and militarized responses. This is not to say that there is no need for policing, enforcement, or intelligence gathering; in fact, giving up on these approaches would not help in tackling the illegal wildlife trade. However, it is important to place these approaches in a wider context and outline what the possible challenges and pitfalls may be. The first is that the quality and availability of data remains a challenge: several agencies and NGOs are repositories of massive amounts of data on illegal wildlife trade (seizure data is just one example). Because of this there is a need for even greater coordination and (safe) information sharing. Second, is the safety and security of those data; some NGOs have developed informant networks and collected large amounts of information, but these data may not be stored or shared in a way that ensures the anonymity and safety of participants. Third, it is important to be aware of the risks associated with intelligence-led techniques. In the drive to get results conservation NGOs can start to perform what are essentially policing duties, which carries the risks of regular policing for NGO staff and their informants, people who are not necessarily trained or skilled in dealing with such challenges. It might be understandable in scenarios where official policing organizations are either overstretched, corrupt, or regarded as an oppressive arm of the state. Even in contexts where policing is more positively regarded and well resourced, there are good reasons to be cautious about the involvement of conservation NGOs to collect and collate intelligence. Fourth, a question remains over whether these approaches address the underlying drivers of illegal wildlife trade, or whether they can deal only with the outcomes and symptoms of more deep-rooted social, political, and economic problems. Therefore, while intelligence-led approaches can be important means of tackling illegal wildlife trade, there are significant challenges and risks associated with them.

6 SECURITY TECHNOLOGIES AND BIODIVERSITY CONSERVATION

While sitting in a sweltering office, I asked a surveillance technology expert about technological approaches in conservation and whether they were important and effective tools. The answer surprised me: "Tech is good but it won't solve everything." The development of new forms of mobile and surveillance technologies are changing the approaches and practices of conservation, rendering them more and more compatible with the security sector. These include camera traps, remote-controlled cameras, more sophisticated remote sensing, systems for tracking rangers, use of artificial intelligence (AI), algorithms, and big data mining. Shifting to these approaches reflects a broader faith in technological solutions for a range of environmental crises, including geoengineering, carbon capture and storage (CCS), blockchain, big data, data visualization, and remote sensing. The rising use of unmanned aerial vehicles (UAVs), more commonly referred to as drones, in conservation is one example. Drones have been hailed as an effective (and low-cost) way of collecting large amounts of data on wildlife populations in remote or difficult terrains. They can also collect information that helps to track rates and patterns of poaching by, for example, recording footage of dead elephants. However, the use of drones is not without controversy. Some organizations have supported the use of drones to collect visual and audio data on people suspected of illegal hunting. NGOs, such as Sea Shepherd, have used them to record and expose instances of illegal killing of seals in Namibia. The drones are provided free of charge by ShadowView, a nonprofit that provides charities and NGOs with unmanned helicopters and planes.[1]

There are also alternative ways that technologies can be harnessed to support environmental justice for marginalized communities. Surveillance technology has been used by marginalized communities and activist groups to expose the illegal activities of more powerful individuals and organizations. This is referred to as "sousveillance" to denote the different, bottom-up power dynamics when compared with more common forms of top-down surveillance by states. While the use of new surveillance technologies is expanding as they become more sophisticated and cheaper, there is little understanding of or debate about who uses them and why, and what the wider implications are.

The central question is, how have these dynamics developed and what kinds of effects do they produce in conservation? From tech giants to conservationists who use their mobile phones to track and report illegal activities, a myriad of actors and everyday activities further drive and embed the integration of conservation and security. Why do conservationists have faith in technological solutions for tackling the wildlife trade? What kinds of technologies are used and what is the role of donors and the private sector in the uptake of these technologies? And finally, are these technological solutions only available to powerful and well-resourced interest groups or can they be used for 'sousveillance' to secure the rights of marginalized groups?

FAITH IN TECH SOLUTIONS

In debates about how to tackle the illegal wildlife trade a great deal of faith has been put in new technologies by governments, conservation NGOs, donors, and (especially) the private sector. The amount of data that can be generated, collected, sifted through, tracked, and analyzed is crucial to tackling the illegal wildlife trade. The generation of "big data" can be important for its more efficient and speedy monitoring of wildlife populations, of threats to wildlife, and of levels and patterns of legal and illegal trade in particular species. Technological solutions are often presented as a cost-effective alternative to more traditional forms of monitoring and enforcement, and they can be used to assist conservationists in everyday tasks and make their jobs easier. It is important not to forget the human side of these stories about the deployment of technological solutions.

The trend toward technological approaches is worth further exploration as it raises serious questions, including who (or what) has access to new data generated, who owns and controls the technology employed in conservation, what implications does the use of technology have for those working in conservation and for communities in areas where the technology is deployed. The use of technology in conservation is not new, of course. There is a long history of mapping out areas to develop national parks, radio collaring animals to track their movements and habits, using traps to monitor wildlife populations, and the use of apps for the wider public to report wildlife crime, including timber trafficking. What is discernible is an increasing use of a range of technologies to provide information, monitoring, surveillance, and communication in response to the recent concerns about poaching and wildlife trafficking.

Technology exists in a social, economic, and political context (see, for example, the ground-breaking work of Bruno Latour and Sheila Jasanoff and, more widely, of Science and Technology Studies). Nevertheless, in conservation and tech circles it is still often presented as politically neutral. The enthusiasm for technological solutions means that their associated challenges are often overlooked.[2] A technology specialist coordinating a number of tech projects aimed to conserve some of the world's most iconic species, encapsulated this thought neatly: "They had these big ideas that somehow technology was a magic bullet that was going to help them solve the wildlife crime challenge. Well it is and it isn't. . . . Tech is helpful, but it can only be an add on . . . there is so much interest in tech that people want to fund drones even if they don't work."[3] Despite the apparent faith in the effectiveness

Demands of civil society for improving timber tracking technology: GPS tags, satellite images, and nationwide CCTV surveillance. Bucharest, Romania. The banner reads, "We want SUMAL, not illegal timber!" SUMAL is a platform developed by the government of Romania, as part of the due diligence requirements of the European Timber Regulation (EUTR), to register every timber transport across the country). Photo credit: George Iordăchescu.

of technology and in other innovations for conservation challenges, they are not magic bullets.

Techno-optimism technology to solve environmental problems is a core theme in environmental debates. Over several decades, techno-optimists, such as Julian Simon, and eco-modernists have argued that human ingenuity and new technology will allow us to avoid the destructive effects of anthropogenic environmental change.[4] However, there are hidden social and environmental costs associated with the development of new surveillance and mobile technologies, especially the ways that minerals are extracted.[5] These minerals are often used in high-tech products, including mobile devices, computers, monitors and televisions, and they have important defense applications in guidance systems, lasers, radar, and sonar.[6] Coltan, for example, is used in mobile technologies. Some of the supply comes from conflict zones in DRC, and is thus referred to as a "conflict mineral": it is mined under poor working conditions, produced at huge social and environmental costs, and has generated significant finance for armed groups. Yet, when minerals are extracted for new technologies to solve environmental problems, the negative social and environmental impacts are often rendered invisible in the debates on their use in tackling the illegal wildlife trade.

It is understandable that conservation would turn to new forms of technology in order to tackle the illegal wildlife trade, since tech, in its many forms, is part of everyday life for people around the world. However, there are significant inequalities in the ability to access technology, the power to decide when and how to use it, and who benefits (financially or otherwise) from the information it generates. Taking a very wide definition of "technology," it can be anything that involves the application of scientific knowledge for practice purposes. But what about the rise in new tech and its links to security in conservation practice? What are the roles of donors, governments, and private philanthropists in supporting the deployment of new technologies in conservation?

Big data is emerging as a key approach to tackling the illegal wildlife trade, and there is an assumption that using big data is effective because it can assist in the prediction and preemption of illegal activities.[7] The promotion of big data solutions to wildlife trafficking reflects wider shifts in the ways that security risks are tackled. Enormous amounts of information about our daily lives are available in open-source format, for example, our movements, who we interact with, what we spend money on, what we invest in. Companies, such as the UK-based Black Dot Solutions, specialize in extracting and combining relevant information from social media, news media, corporate data, web domain information, and search engine content.[8] Such information can be used to build up pictures of individuals identified as possible suspects in poaching and trafficking, as well as developing analyses of emerging risks and threats.

Artificial intelligence has been explored by some researchers and conservation NGOs as a means of tackling the illegal wildlife trade in the online and real worlds. Machine learning centers on algorithms that can learn from data without ongoing human guidance. They rely on the development of artificial neural networks that allow a machine to use an algorithm to perform a task repeatedly, learn from the results, and change the algorithm slightly each time. Deep-learning algorithms can be used to analyze language and classify online images, which includes the capacity to identify individual objects within those images.[9] Di Minin et al. argue that the development of machine learning and algorithms provide a way forward for detecting, tracking, and curtailing the illegal wildlife trade in the online world.[10] They suggest that applying these techniques to social media data allows human behavior to be investigated on an unprecedented scale, yet these methods are rarely used in conservation circles. Di Minin et al. further call for social media platforms to allow the use of deep-learning algorithms and to pass relevant information on to enforcement agencies.

As with all technologies, it is important to place the expanding use of technology and data within its social, political, and economic context. As conservationists call for greater use of new techniques from artificial intelligence to tackle the illegal wildlife trade, lessons can be learned from critical security studies about the use of such data and techniques. In the conservation community it is acknowledged that new technologies come with new risks and require better ethical oversight.[11] This is illustrated by focusing on how algorithms are produced. The algorithm is the product of a range of initial human decisions: which training dataset is used, where to set the threshold for sensitivity, how to operationalize models, and so on. The algorithm is in itself integrally part of human thinking and action. Artificial intelligence, machine learning, and neural network computation are radically reshaping practices of securitization and instituting new logics for the governing of populations.[12] The proliferation of algorithmic techniques in the gathering of intelligence data from a vast range of sources inevitably produces political and ethical challenges, but these have not been adequately addressed. Proponents of big data approaches suggest that the collection and analysis of huge amounts of personal data offers the ability to engage in prediction and preemption and thus prevent criminal activity. But this approach in itself raises important challenges, which have not been fully thought through in the conservation sector.

Rules-based algorithms determine who and what surfaces for security analysts (or conservationists in this case). But the shift toward non-rules-based machine learning, via the use of deep learning algorithms, produces new forms of political authority, allowing the algorithms to bring new worlds of understanding into being, for better or worse.[13] In the world of wildlife trafficking, it opens the possibility of using deep leaning algorithms to identify particular individuals or groups for targeted intervention to preempt or prevent poaching or trafficking. It appears a very attractive option, but it also has potential to target groups and individuals for interventions by authorities in state agencies, private security companies, and conservation NGOs seeking to protect wildlife.

Intelligence agencies and conservation NGOs have used big data mining techniques to identify potential poaching hotspots and those involved in organizing poaching and trafficking. Intelligence agencies are using all data available to them, most of it from the unclassified world. For example, information from banks about large transfers of money from suspect individuals can help intelligence agencies predict whether an order has been given to poach or buy illegal wildlife products. Intelligence agencies can then alert the relevant national authorities to put more protection in place for certain wildlife populations in particularly vulnerable locations, at least in theory. Furthermore, intelligence-based approaches are in the business of prediction and preemption. For example, setting the terms and definitions for the initial algorithms is in itself an act of inclusion and exclusion, which will inevitably focus attention on some activities but ignore others. It is not necessarily a perfect tool for tracing all illegal wildlife trade—the trade itself is the product of global social, political, and economic dynamics which are not amenable to simple or quick solutions.

The use of AI, machine learning, and algorithms can have unforeseen consequences. Tracking online activities with algorithms affords those using them enhanced powers to draw conclusions from vast amounts of private data. For example, Google and ZSL are developing image recognition technology to democratize the ability of individuals and organizations, with more limited experience and capacity, to analyze data and report suspicious activity to wildlife law enforcement agencies.[14]

However, there are important ethical issues, as we can see from research in other fields on the use of social media platforms. For example, Lally's study of how Reddit was used to track down the Boston Marathon bomber showed that with big data, it is possible to engage in wide-scale surveillance of populations, thus to track individual subjects and make important inferences about their thoughts, actions, and motives. In the process people can be wrongly identified as engaging in illegal behavior and may find themselves marked as suspicious individuals,[15] which can lead to profiling, enhanced surveillance, or unwarranted and unwelcome contact with, or intrusion from, law enforcement agencies. There is much the conservation sector can learn from the emerging critiques in security studies about machine learning, use of algorithms, and big data.[16] Data collection is not an apolitical, technical tool that can used to enforce the law in a neutral way. The ability to access, collate, and sort private data streams, which follow us through our everyday lives, accords a great deal of power to those collecting and using such data.

In conservation, the analysis of big data has been used to identify patterns and perpetrators of the online illegal wildlife trade. E-commerce is rapidly developing in value and volume. The WCO acknowledges that cross-border e-commerce is increasing because of advancing internet technology, economic development in the global South, expansion of express delivery services, and mobility of the labor force. According to estimates made by UNCTAD in 2013, business-to-business e-commerce was valued at about US$15 trillion and business-to-consumer e-commerce at more than US$1 trillion, with the latter growing faster in the last few years.[17] In recognition of this, the Global Transparency

Initiative (GTI) launched its Digital Dangers Project, a collaboration with INTERPOL and UNODC, to provide policy advice on how to disrupt online illegal wildlife trade.[18] The internet has had an increasingly significant role in legal and illegal wildlife trading.[19] New online technologies, including social media, provide new forums for traders and consumers to connect with one another. Several online platforms have had to respond to illegal trafficking and have introduced measures to regulate or ban wildlife trading. In 2008 the online marketplace Taobao banned the sale of species included in the Wildlife Protection Law in China, and Alibaba, an e-commerce site, banned the posting of a range of protected animals, plants, and their derivatives. Ivory trading was banned by eBay in 2009, and in 2013 Etsy banned the sale of ivory and all other products made from endangered species. These bans on online trading platforms meant that the illegal wildlife trade shifted to social media platforms, including WeChat and WhatsApp.[20]

The online trade in ivory is a good illustration because it is very challenging to identify and track. In their analysis of ivory trade in four EU countries, Alfino and Roberts demonstrate that online traders extensively use code words for ivory in order to evade detection. This practice created an additional layer of complexity for law enforcement, as well as for eBay in its attempts to close ivory trading on the site. Because it was difficult to keep pace with the use of different code words for ivory, trading continued on the platform despite commitments to tackle online illegal wildlife trading.[21] The proliferation of more sophisticated social media platforms has also facilitated new ways of trafficking wildlife products, and such trading presents additional challenges when compared with the major internet auction sites like eBay.[22]

Conservation NGOs and the law enforcement community have recognized online platforms as a key method of trading wildlife. For example, Project Web, a collaboration between INTERPOL and IFAW, identified the internet as an important means of trading ivory illegally. However, CITES—as a well as national governments, the internet, and social media companies—is ill-equipped to deal with online trading, and legal frameworks have not kept pace with its development.[23] IFAW, for example, identified cybercrime as a critically important field to investigate in order to tackle the illegal wildlife trade. In 2014 IFAW spent six weeks monitoring the online trade in 280 online market places, covering species listed on CITES Appendices 1 and 2 (which means trade is completely banned or is strictly controlled). The investigative team found a total of 33,006 endangered wildlife, and wildlife parts and products, available for sale in 9,482 advertisements. Ivory, reptiles, and birds were the most advertised items, and they had an estimated value of at least US$10,708,137.[24] The researchers recognized the limitations of their investigation, including the difficulty of identifying species from online descriptions and images and of detecting trading scams in which fake or nonexistent ivory was offered to buyers.[25] They pointed to the challenges of working with the law enforcement community in places where corruption is an issue. The team did not hand over its intelligence to the authorities in situations where it might be used for corrupt purposes. It also prioritized handing over intelligence to agencies defined as having adequate enforcement capacity.[26]

China's market in illegal wildlife products is roughly estimated at US$10 billion per year.[27] China has the world's biggest online community, at more than 641 million internet users (2015 figures), and has also seen rapid growth of e-commerce. This includes illegal wildlife trade on the internet. Traders can use code words for illegal products and hide their real identities, making it harder to prevent or detect illegal wildlife trade.[28] In China TRAFFIC worked with WeChat,[29] one of the world's largest social media organizations, and with Ali Baba,[30] an online trading platform, to prevent illegal wildlife trade. In 2012 IFAW also persuaded the online forum Baidu to shut down thirteen forums that discussed the illegal wildlife trade.[31]

In 2012 TRAFFIC-International began tracking fifteen websites in China for trade in ivory, rhino horn, tiger bone, hawksbill shells, and pangolin scales. The team had to constantly keep abreast of and update code word searches because dealers were very adept at changing them to evade detection. From 2014 the team routinely used machine learning rather than manual searches to monitor illegal online trade, which significantly reduced the time and effort needed to trace the trade. Monitoring and reporting advertisements for illegal wildlife products did reduce the number of adverts from a high of 4,000 per month in 2012 to 1,500 per month in 2015. However, TRAFFIC-International also point out that such detection and reporting, even while using machine learning to speed up and expand the scope of searches, means that large numbers of adverts are missed because they use different code words and methods of trading.[32] Indeed, a report of the Global Transparency Initiative into online illegal wildlife trade concluded that to date attempts to shut down these markets have failed, and at best they have only managed to temporarily disrupt the trade.[33]

In June 2018 IFAW and INTERPOL held a workshop on cyber-enabled wildlife crime. The purpose was to explore the ways that cybercrime intersected with illegal wildlife trade, and it developed a joint call to action for the UK government London Conference on the Illegal Wildlife Trade in October 2018. One of the aims was to create a single online portal to which NGOs, law enforcement agencies, academics, policy makers, and the private sector can upload information. It would then be used to share knowledge about perpetrators and potential perpetrators of online wildlife crime. The aim, ultimately, is to use the "data acquired during monitoring activities to establish strategies regarding enforcement, capacity building and public awareness."[34] Therefore, the call to action is explicit in its aims of moving from detection and monitoring to use of big data for targeted enforcement.

The above examples also reflect the level of power and autonomy conservation NGOs can have in tackling the illegal wildlife trade. NGOs can generate their own intelligence, conduct their own investigations, and develop their own means of dealing with the illegal wildlife trade. But how are these strategies shaped by the different kinds of tech that are used in conservation?

TYPES OF TECH

The engagement of conservationists with forms of monitoring and surveillance using new digital technologies reflects much broader shifts in wider global society. As Sandbrook, Luque-Lora, and Adams argue, the surveillance in conservation contexts follows the rise of surveillance in wider society, particularly in security and policing by state and private agents. Furthermore, the development of intensive surveillance regimes are integral to new forms of conservation governance, such as enhanced policing of protected areas to counteract poaching.[35] As with other security-oriented approaches in conservation, these new developments have been supported and expanded by the entry of new stakeholders in conservation, and they have produced new practices on the ground. As one conservation tech specialist commented, "Now there are other organizations which were not focused on wildlife, not conservation organizations, like C4ADS, think tanks, founded originally doing defense-related research but now they have quite a strong presence, they do reports on rhino horn and ivory trade. They bring new techniques, they do what TRAFFIC does but bring a whole bunch of different kinds of methods and analysis."[36]

Advances in surveillance technologies and falling prices have allowed widespread adoption of digital surveillance technologies in conservation. Furthermore, the faith in technological solutions has attracted, and then facilitated, the two-way exchanges of research and technologies used for security, as well as for conservation. For example, there are clear crossovers between the two sectors, as one tech specialist pointed out: "One of the computer sciences labs at the University of Maryland, involved in analyses of social media, looking at Afghanistan, Iraq, Pakistan, and India, . . . got interested in this wildlife trafficking issue. There is a unit at UCLA that is computational conservation, they do game theory . . . how do you deploy your resources in a way that look random to the bad guys but is not random."[37]

Technological solutions inevitably have their limits, and there is some recognition of this in conservation circles. Those experienced in anti-poaching acknowledge that while technology can help it is not a substitute for the basics, such as having a sufficient density of well-trained, adequately equipped, and strongly motivated field rangers. As one conservation technology specialist remarked, "Useful things can be done with tech in terms of game theoretical algorithms or optimising your patrols and searches and law enforcement activities, also amazing toys, but none of it is really very useful if you don't have a solid foundation for law enforcement."[38] In short, new technologies still require personnel on the ground to make them effective, which is a well-documented problem in the case of technological approaches to warfare.[39]

Aerial surveillance technologies are important in this fast-developing approach. Conservationists have been especially concerned with how to make aerial surveillance work for ecological monitoring and law enforcement. However, there are wider implications that also need to be considered, including the ways these techniques can restructure human relationships with nature. The deployment of aerial technologies in conservation has the power to reconfigure socio-natural and socio-spatial relations. They can be a central factor in processes of territorialization and in controlling access to and uses of natural resources.[40] The expansion of aerial technologies is not just about the development of the capacity of technology itself, it has been extended as part of wider responses to poaching by conservation agencies. In the case of Kruger National Park, the deployment of UAVs, helicopters, and remote sensing technologies is specifically to enhance security and the capacity of anti-poaching patrols. The anti-poaching units are able to obtain up-to-date information on the precise locations of elephant herds, rhinos, and suspected poachers. Being able to do this enhances their capacity to predict where poachers might strike next, and it allows for a much greater level of knowledge about the park and the distribution of wildlife within it.[41]

One of the areas that has received little attention to date is the effect that drone use may have on the wildlife itself, in terms of behaviors, movements, and stress levels. Amid some growing concerns about drone use for wildlife monitoring and for law enforcement, Mulero-Pázmány et al. undertook a systematic review of research papers that contained data gathered by UAVs in order to understand the effects on wildlife.[42] They found that wildlife reactions depended on a combination of UAV flight pattern, engine type, and size of aircraft, as well as on the characteristics of the animals themselves (type of animal, life-history stage, and level of aggregation). They concluded that when drones were used to target particular movements of animals, larger and noisier drones evoked the strongest reactions. The effects on animal behavior were much more pronounced in birds, in part because birds may have reacted to the drone as a predator.

The use of drones raises a much wider issue about the development of technological solutions to tackle the illegal wildlife trade. These interventions can fundamentally change the ways that animals behave and can reconfigure relationships between animals and human communities. For example, if drones are used to deter elephants from crossing boundaries between national parks and agricultural land, it fundamentally affects the behavior of those elephants and their interactions with human communities. While Mulero-Pázmány et al. focused on how drone use changed wildlife behaviors, they acknowledged that we do not know enough about how the deployment of these technologies may be creating physiological changes, including increased levels of stress, in wildlife populations.[43] There are clear animal welfare concerns in the deployment of these new technologies, and there is a need to reflect more fully on their impacts and then develop instruments geared toward animals' welfare needs in any new conservation initiative.[44] To date the welfare concerns have revolved around stopping poaching and trafficking, rather than the welfare implications of new technological interventions to protect animals themselves.

There are also potential ethical, political, economic, and social effects of reliance on technology in conservation. For example, the American Association of Zoos and Aquariums supports Zoohackathon events, with the tagline "coding to end wildlife trafficking."[45] The idea of a Zoohackathon is that tech experts are invited to solve problems presented by experts on wildlife trafficking. At the end of a Zoohackathon, the tech specialists pitch their ideas to the wildlife conservation community. In 2017 the winner at the London meeting was Team ODINN, which developed a product that intelligently processed images from camera traps in the field, so that information could be passed to rangers on the ground to alert them to the presence of particular animals or people. The purpose was to allow rangers to identify poaching threats in real time and respond more immediately.[46] This is an example of the innovations that technological solutions can bring in to conservation.[47] However, the use of machine learning to analyze images much more rapidly so as to identify the presence of people is not a politically neutral approach. The positive promotion of ODINN sidestepped the complex politics of collecting, analyzing, and using images of people captured for law enforcement activities, whether intentionally or unintentionally.

There are controversies over the ways that surveillance and monitoring technologies are used in conservation. For example, a recent study by Sandbrook, Luque-Lora, and Adams indicates that camera traps are now widely used in conservation research and practice, but the wider issues around their use and how people on the ground respond to them is not adequately understood.[48] They suggest that the extension of camera trap use can be regarded as an important tool in the expanding regime of conservation surveillance precisely because they unintentionally capture images of people, or *human bycatch,* as they put it. Ninety percent of the users of camera traps that they surveyed reported that the cameras recorded images of people. These unintentionally gathered images were then, in some cases, diverted to enhance law enforcement, creating greater conflicts with conservation initiatives. Further they argue that conservationists can be in breach of commitment to do no harm. Over 75 percent of their respondents reported objections to or direct interference with camera traps, which indicates that people can oppose their deployment because they are concerned about what data is being collected and how it might be used.[49]

The resistance to new forms of surveillance in conservation can be the result of a lack of information about the purposes of the cameras. This was the case in Terai-Arc National Park in Nepal, where cameras were vandalized. People in the area were not adequately informed about the purpose of the traps and of how the images of illegal activity would be used.[50] In one of the most high-profile examples, nine Iranian conservationists from

Camera trap, Indonesia. Photo credit: Laure Joanny.

the Persian Wildlife Heritage Foundation were arrested in 2018 and five were charged with "corruption on earth," which can carry the death sentence. They had been using camera traps to monitor endangered Asiatic cheetah populations, but the Iranian Revolutionary Guard accused them of espionage, of using the traps to monitor Iran's ballistic missile program and collect sensitive data for hostile nations.[51] By early 2019, their case had attracted international attention and support, and Erik Solheim from the United Nations Environment Program (UNEP) expressed official concerns to the Iranian regime.[52]

There are several hidden challenges with using new forms of monitoring and surveillance technology for conservation purposes, not least that it can mean conservation NGOs can hold much more sensitive information on the movements of animals and peoples than governments do. But this is not to suggest that there are no beneficial effects for people living in or using areas where camera traps are deployed. If the images are shared with people, then they can increase support for conservation initiatives.[53] Further, camera trap images can also be used by groups of activists and local communities to document and report illegal activity in their areas. There can be significant risks for conservationists too.

REMOTE INTELLIGENCE GATHERING

Within conservation circles there is growing interest in remote intelligence gathering and its capacity to offer a range of new tools to count and monitor animal populations. Three main methods are used: satellite imagery, images from drones, and imagery from light aircraft. The growing availability and use of drones has been heralded as revolutionary in ecology.[54] Drones offer the possibility of engaging in surveillance and monitoring over much larger areas of land and sea than can be covered by human patrols (such as by rangers in protected areas).[55] However, a review of these approaches by Hollings et al. pointed out that they have limitations, including relatively low accuracy of automated detection technology across large-scale landscapes, false detections, and the costs of high-resolution imagery.[56]

Testing a small conservation drone (pictured on the ground), Indonesia. Photo credit: Laure Joanny.

The use of new technologies in conservation has some major financial backers, and the uptake of technology is indicative of the power of donors and the private sector to shape policies to deal with the illegal wildlife trade. For example, Google's Global Impact Awards funded US$23 million to develop technology uptake in areas that included conservation and humanitarianism. In 2012 Google granted a US$5 million Global Impact Award to WWF to develop the Wildlife Crime Technology Project, in order to explore and develop new technological approaches to countering poaching and trafficking. These initiatives included the use of drones and camera traps for poacher detection. In 2012–14 Phase One focused on the development of drones and of Spatial Monitoring and Reporting Tool (SMART) ranger patrol analysis technology.[57] Such funding has been used by the providers, in this case Google, to move into the conservation sector. This scenario is not necessarily problematic, but it does mean that Google can benefit from its investment in reputational terms and by developing new approaches that rely on products provided by the company itself. Furthermore, such initiatives bring modes of thinking and practice from the tech sector to conservation.

In response to the proliferation of new technology projects, often funded and run by organizations from outside, several African governments began to develop legislation on the use of drones, and governments started to resist adoption of SMART.[58] The development of new technologies has been so fast that it has been difficult for legislation to keep up. Nevertheless, the government of Kenya did not approve the use of drones for civil, nonmilitary purposes until 2018 when the Kenya Civil Aviation Authority (KCAA) developed new guidelines for their use, which meant they could not be used in national parks or near sensitive facilities, such as airports.[59] As the use of drones became more restricted, Phase Two of the Wildlife Crime Technology Project focused on using radar and acoustic approaches to detecting gunshots and poacher movements. In 2015 WWF field-tested prototype thermal cameras and machine learning, with the aim of identifying humans, so that they could trigger automated alerts about potential poachers for rangers.[60] Further, the lowering costs and easy availability of drones has fostered their proliferation in conservation practices, for activities as diverse as wildlife counting and fire prevention.[61] Drones raise wider questions about the ways that technologies can be transferred from the military sector to conservation,[62] and about the limitations of such transfers, not least that national governments may not welcome drone use by NGOs. It is critically important to place this in a wider context in order to understand the potential challenges they can produce.

The growth in drone use has some limitations that are worth exploring further. For example, they need to refuel, the imagery produced is at a lower resolution because they can carry only lightweight (payload) machinery, they are affected by the weather (especially wind), they can be expensive in terms of labor because operators have to be trained, and there can be legal restrictions on the use of drones in public spaces.[63] The backers of drones often have often not thought through the logistics of if they will be able to operate in particular environments. As one conservationist remarked there was a sense of "Oh, you have to get drones." "Yes, okay. Did you think about your energy source in the Amazon or in DRC?"[64]

Certainly, drones can provide information that is otherwise difficult to obtain, but there is a need for wider reflection on such questions as what data they can and should collect, who should be able to access collected data and for what purpose, and even whether the conservation drones could also be deployed for dual purposes (such as collecting information of interest to security services, law enforcement, or other arms of the state). Furthermore, as Sandbrook points out, their deployment for conservation raises numerous issues of safety, privacy, and psychological well-being.[65] Drones open up new dimensions in recording, sharing, and analyzing images.[66] They have the potential to identify and track individuals (through facial recognition software), which of course can be important for law enforcement, but it can also be used for the exercise of power through a state or private sector security apparatus over more vulnerable, marginalized or resistant populations.

Drones are presented as a cheap and effective contributor to anti-poaching because they can cover much larger areas in a shorter time than teams of rangers traveling on foot or even in vehicles. However, the information they provide is only useful if it can be integrated with the activities of rangers on the ground. In order to be useable and effective, people on the ground need to be ready for immediate deployment if the drone information identifies potential poachers. Furthermore, although drones can be useful in open areas or urban environments, in Kruger National Park in South Africa they were discontinued after a year-long trial because they were unable to "see" through thick bush.[67] While they are often regarded as a cheap and easy approach, drones can be expensive to operate and require specialized training to fly them effectively. For example, the estimated cost of running a drone program in one small protected area in Kwa-Zulu Natal Province, South Africa, was US$500,000 a year. The province has twenty-three other parks and reserves. If drones were deployed in all of them it would cost a total of US$12 million—and if poachers were detected by the drone, a highly trained and well-led ranger on the ground would be needed to have a chance of intercepting them.[68] There is a great deal of excitement about the potential of these new technologies, but they come with additional challenges, not least their costs and reliance on staff on the ground. The allocation of often scarce funding to drones inevitably takes away from financing other activities or equipment.

Drone use by nonstate actors has also been curtailed by national governments—across the world—as states have tightened up legislations on their use by private citizens, NGOs, and the private sector. The development of regulations governing drone use has shaped the level of uptake in conservation. In one example of an actor from the security sector becoming more integrated with conservation activities, Thomas Snitch, a computing specialist, aimed to apply the lessons from a model he developed to protect US troops in Iraq to tackle poaching in Africa.[69] The aim was to enhance conservation and develop evidence bases for prosecutions in cases of wildlife crime. Snitch's team at the University of Maryland developed a model using data from satellites, drones, and analytics to predict where rhinos might be located on full moon nights (when most poaching takes place). Ranger teams could then be deployed near where the rhinos were most likely to be, and the drones could be deployed further away to detect potential poachers as they moved through protected areas.[70] The purpose was to collect data for use in prosecutions, so that prosecutors could have confidence that data had not been tampered with by corrupt officials who might be involved in poaching and trafficking themselves. However, the growing restrictions on drone use by African governments slowed the uptake of the system he developed.

There is also much to learn from the longer-term use of drones, especially their limitations, in the military and security sectors. The use of drones in warfare as instruments of surveillance and as weapons, has been widely debated in the disciplines of security studies, international relations, and human geography.[71] Similarly their adoption into routine police surveillance and control also holds useful lessons.[72] In terms of western military strategy, the development of drone warfare is part of a wider shift toward light footprint military intervention, which relies on military to military (M2M) training, use of remote technologies and surveillance, private security contractors, special forces, and precision bombing. The nature of this open-ended warfare, in which people are harmed and killed, is rendered obscure, unaccounted for, and elusive.[73] However, conservationists could learn from these debates in security studies. The use of drones in conservation in some circumstances can be integrated with the broader military strategies of the world's major powers, especially in Africa. Drones can also reproduce and deepen problematic and exclusionary imaginaries of wildlife and landscape in Africa.[74] As both Millner and Ybarra argue in their work on Belize and Guatemala, respectively, the incorporation of drones into conservation planning and management can be used to justify increased military presence, including the deployment of wider surveillance technologies as well as stop-and-search strategies, and as a result conservation frameworks have facilitated the containment, surveillance, and management of populations defined as "risky."[75] In short, new technologies such as drones can support and reinforce existing power dynamics and can even bring into being new power relations between people and wildlife.

The political ecology of the growing faith in technological solutions is an important but underacknowledged issue. The turn toward technology can divert investment and donor funds away from developing better working conditions, equipment, and wages for rangers. Further, it opens new opportunities for the private sector to become involved in conservation enforcement and new product development. As such, the growth of a new industry around wildlife protection and anti-trafficking produces vested interests in developing new sources of accumulation. As Massé and Lunstrum show, in South Africa the development of new securitized spaces for anti-poaching developed opportunities for accumulation by private sector actors, which profited from conservation law enforcement.[76]

PHILANTHROPY AND THE TECH GIANTS

Donors have supported the development of new technological solutions for tackling the illegal wildlife trade. USAID is an important donor in this field via its Wildlife Tech Challenge Fund, which encourages new players from the tech field into conservation. The Wildlife Tech Challenge Fund was implemented by a consulting firm, Integra LLC. Price Waterhouse Coopers provided oversight, and National Geographic, TRAFFIC, and Smithsonian Institution were all partners in the Challenge. The panel of judges were Crawford Allen (director of TRAFFIC—USA), Bryan Christy (a journalist for *National Geographic*), and Mahendra Shrestha (from Smithsonian Institution and the Global Tiger Forum). The idea behind USAID Wildlife Tech Challenge was to provide seed money to develop innovative technological solutions to tackle urgent challenges for wildlife conservation. The judges drew up a list of forty-four finalists; sixteen were awarded prizes of US$10,000, and the winners were also allowed to compete for four additional grand prizes of up to US$500,000 each.[77] The seed funding enabled the ideas to develop to a point where prize winners could enter into partnerships with funders to refine and scale up their innovations. The Wildlife Tech Challenge team also helped with communications to create a buzz around potential tech solutions. From the outset, the Wildlife Tech Challenge aimed to bring in new groups, including smaller entities, even one-person operations that needed access to some funding and support to develop their ideas. This was because there was a sense that the big conservation entities, such as WWF-International, tended to work with the same organizations again and again.[78]

Philanthropic foundations and private individuals are also key financial supporters of technological approaches to tackling the illegal wildlife trade. One important factor in this is that several of these private donors have developed their fortunes in the tech sector themselves, and they seek to develop and apply lessons from that sector. There are many others, in addition to Google and the Google Global Impact Awards. For example, the Wildlife Conservation Network (WCN), an NGO based in Los Altos, in California's Silicon Valley, was founded by IT entrepreneurs and executives from such companies as Cisco Systems and Yahoo. The WCN is explicit about applying Silicon Valley venture capitalism to conservation philanthropy; it serves as funder and mentor for more applied conservation organizations.[79] Vulcan Inc. Philanthropy, the foundation of the family of Paul G. Allen (a co-founder of Microsoft), is one of the most publicly active funders for tech in conservation arena. The World Bank analysis of donor funding for tackling illegal wildlife trade indicates that it committed US$19 million between 2010 and 2016.[80] Among other things, the foundation funded the Great Elephant Census, to record elephant numbers across forty countries, and it supported the development of techniques for DNA coding to identify the source of elephant ivory.[81] Such funding obviously has significant capacity to shape responses to the illegal wildlife trade. Philanthropists can be nimble and more responsive to emergency situations than other kinds of donor; they can provide speedy support to priority areas and can offer assistance when the need is urgent in a way that other donors, who are held to funding cycles or need the agreement and authorization from several different sources first, cannot. Philanthropists often come from a business culture of setting specific goals and expecting rapid, clear, and trackable results. On the one hand, it can mean clear, measurable targets, which can enhance the ability of conservationists to tackle the illegal wildlife trade. Yet, on the other hand, as some conservationists reported to me, it can also translate into pressure on conservation professionals on the ground to redirect their efforts from activities that are not so easily demonstrated in terms of measurable outcomes; the examples cited were pressures to increase the number of seizures of trafficked goods, to obtain more arrests of poachers, or, in some cases, to adopt a culture of more aggressive anti-poaching in order to get the results their philanthropic donors required.[82] The interest and financial support from private philanthropic foundations in technological solutions has further expanded the involvement of the private sector in conservation—which has been important in conservation, especially in the last twenty years.[83] The confluence of philanthropy, new technologies, and security tackling the illegal wildlife trade opens up new markets in conservation, from which the private sector can benefit substantially. But these new frontiers for accumulation can mean diverting funds to technological solutions and away from everyday activities, such as ecological monitoring, or basic equipment, such as adequate clothing.

SOUSVEILLANCE, TECHNOLOGY, AND ACTIVISTS

The focus of this chapter thus far has been on the use of technology by more powerful actors in conservation. There are, however, ways that technologies can be deployed by more marginalized and vulnerable groups, and this strategy is gaining ground but remains under researched.[84] New technologies are embraced and deployed by less powerful groups seeking to expose wrongdoing or injustices at the hands of corporations, political and economic elites, and corrupt government officials. It is sometimes referred to as *sousveillance,* or the use of surveillance technology to monitor more powerful authorities from below. In practice it differs from the ways that very powerful actors have funded and facilitated the entry of technology companies into conservation and its linkup with the security sector. Using a political ecology lens can cast light on the struggle of more marginalized peoples in establishing and maintaining their own rights. It also reveals that there are alternative approaches to the use of technology, uses that can be participatory and support claims for rights to land and resources. One conservationist working with forest dependent communities stated: "Last year, a soy farmer wanted to deforest a piece of land, deforest again. They already had conflicts over other pieces of forest. He started doing it with the bulldozer. The group, they arrived with the camera, filming and taking pictures. Then they stopped, they took these pictures to the public prosecutor. There was a hearing."[85]

This is especially important where communities have a trust relationship with law enforcement or can link up with global NGOs pressing for their rights to be recognized and protected. When there is trust between law enforcement actors and (leaders of) forest communities, GPS pictures and other intelligence from the ground can be communicated to law enforcement and criminal justice agencies.[86]

There are multiple examples of the ways that groups can use and reconfigure technologies to establish their own rights over land, forests, or wildlife, as well as to expose the illegal activities of governments, logging companies, and poaching networks. Cameras, mobile phones, GPS, and other technologies can be used by communities and activists who seek to expose illegal activity by powerful actors, or to map and establish rights over land and resources. There has been an ongoing debate on how counter-mapping by vulnerable or marginalized communities can help to establish their rights over land. In the 1990s radical counter-mapping arose as a means for subaltern communities to use novel, inexpensive forms of geo-referencing to produce maps of their own communities. One well-documented example is the Maya Mapping Project in Belize, which produced the *Maya Atlas*. This map, produced by communities in the Toledo Maya Cultural Council, aimed to articulate the argument for indigenous rights to an international audience. However, the map, as with all maps, was an expression of particular interests within heterogeneous Mayan communities. For example, the *Atlas* showed spaces of men's work (hunting, cultivation) but not of women's work (primarily in the home).[87] Those early mapping exercises have now been updated with the use of new technologies. For example, in the case of the Maya Biosphere Reserve in Guatemala, drone technologies have been introduced by community organizations, rather than by the state or conservation organizations, as part of their own efforts to protect areas under community management from external threats. The communities have organized a monitoring network using drones to collect data—photographs, video footage, and biomass measurements—which are combined to create maps and graphics that tell stories about forest change and protection.[88] In the Brazilian Amazon, Tim Boekhout van Solinge has drawn on the networks he previously established as part of his academic research on criminology to create a system for supporting residents in reporting forest crime. They have begun a community forest watch using GPS and cameras to send images and location information about illegal logging and land grabs to trusted law enforcement officials.[89] These methods can be used to demarcate land for land claims and provide photographic evidence on infringements.

In one of the best documented examples of alternative uses of technology in conservation, Rainforest Foundation UK has been actively supporting the use of technology to allow communities to document their claims to land and to expose illegal logging in their areas. It has developed ForestLink, a real-time monitoring device that allows communities in Peru, Cameroon, Ghana, and DRC to send evidence of illegalities to law enforcement authorities, even in areas with no mobile connectivity.[90] Since 2011 it has also developed and supported the MappingForRights program, which allows communities to map their area, bringing into view the often-invisible lives and activities of indigenous forest-dependent peoples. The MappingForRights program has also included legal training for communities, to allow them to press for legal recognition of their rights.[91] As these examples show, technology used in conservation has a range of applications, and it is not always the case that more powerful actors employ it to further their own interests. Instead, the application of newly developed technologies can enhance community participation. It could be more effectively harnessed and deployed to help support communities in tackling the illegal wildlife trade, and to allow them to expose the detrimental activities of logging companies, poaching networks, corrupt officials, conservation law enforcement, and NGOs.[92] Similarly, BaYaka communities in the Republic of Congo use mobile phones and GPS devices provided by the Extreme Citizen Science (ExCiteS) team at University College London to map and document the changes in the forests they depend on, thus telling the story of how the forest has been opened up to development agencies, loggers, conservation NGOs, and urban elites, pushing forest-dependent peoples, BaYaka in this case, further into the margins, into low-paid work, increasing stress, and weakened social and familial ties.[93] These examples indicate that technology can be harnessed to protect the rights, lands, and interests of marginalized peoples if they are introduced, managed, and used in ways that are beneficial to them, rather than as a means through which more powerful actors segregate wildlife and people or enforce laws that some communities regard as unjust and a threat to their survival. There are growing demands to decolonize conservation and to recognize the human and land rights of indigenous peoples. Several organizations, including Survival International, Rainforest Foundation UK, and the ICCA Consortium,[94] promote the message that indigenous peoples are central to conservation and play a vital role in protection of the environment, arguing that forests and wildlife have fared better under indigenous peoples.

New forms of mobile and surveillance technologies are rapidly reconfiguring conservation. These technologies include camera traps, remote-controlled cameras, drones, remote sensing systems, and surveillance of mobile phone communication. In many respects the shift toward more technological solutions for conservation has facilitated its compatibility with the security sector, but this is only part of the story because technologies can also be harnessed by vulnerable groups to protect or press for recognition of their rights and roles in conservation.

It is important to draw attention to the political economy of the shift toward technological solutions: who benefits financially and who is empowered by the entry of new forms of technology into conservation? By contrast, who or what is disadvantaged or disempowered? The faith in technology certainly provides more opportunities for the private sector to enter conservation practice and to profit from it. As is apparent from the involvement of philanthropists, the tech sector builds on the earlier neoliberal phase of conservation. Further, the deployment of new technologies can change the relationship between humans and nature—and the resistance to camera traps indicate a change in the relationships between nature and the human communities that have been traveling through landscapes. Suddenly, their activities are rendered visible to a wider community of conservationists and surveillance specialists, with all the challenges that produces for personal privacy, anonymity, and security.

Furthermore, it is important for the conservation sector to learn from the pitfalls of reliance on big data, algorithms, and AI in other arenas. The ways that algorithms are developed out of a complexity of human decisions, including prejudices, is key. The use of algorithms is not politically neutral. Like all information is it embedded in, and shaped by, wider power structures and a social, political, and economic context. As with other new approaches in conservation, there can be positive uses of such technologies, as the examples of sousveillance reveal. The availability and affordability of some technologies has provided marginalized groups the opportunity to expose the wrongdoings of more powerful actors.

7 MILITARY-SECURITY-CONSERVATION NEXUS

One of the noticeable shifts in conservation is the growing collaboration between militaries and conservationists. It is discernible in places that are not conflict zones but where there are struggles over the environment, wildlife in particular, and it is changing the ways we think about responses to the illegal wildlife trade. Such collaborations are not a straightforward importation of military ideas, techniques, and practices. Rather, engagement with the illegal wildlife trade has enabled militaries to develop new approaches and has allowed them to present themselves in new ways to a wider public. It is a two-way exchange of ideas, approaches, and practices between militaries and conservationists.

In this chapter I use the term "military" in a broad sense, encompassing the military forces (air, sea, and land) of nation-states, international intervention forces, such as UN Peace Keeping Operations (UNPKOs), and private military companies (PMCs). Standing armies have had a long relationship with conservation, and the current phase builds on those earlier historical engagements. Conservation practice more generally has seen a rise in the use of PMCs to deliver anti-poaching training or even to engage directly in enforcement. But what unites these examples? Are they part of a more general and longer standing pattern of integration between conservation and security?

HISTORICAL ENGAGEMENTS

The current intersections between militaries and conservation build on a history of engagement between the two sectors. It is important not to overstate the newness of the growing collaborations with military actors. There are precedents, which help to explain why the uptake of a security-oriented approach to tackling the illegal wildlife trade has been so readily accepted. For example, under colonial rule the bulk of European staff in state conservation agencies in Tanganyika were demobilized soldiers from World War Two. They had no training or background in conservation, but they did have training in military operations and weapons use, as well as time on their hands.[1] Similarly, in the late 1980s the KAS Enterprises team involved in Operation Lock retrained several members of the notorious police anti-insurgency unit, the Koevoet, in Namibia as game wardens.[2] In Guatemala, the national army has been deployed in conservation roles to manage the Maya Biosphere Reserve.[3] It should not be surprising, since there are clear overlaps in the skills needed in some conservation activities and in a range of military roles. These include ability to work in remote or hostile environments with little infrastructural support, working in small teams, and understanding landscapes and how to navigate them.

In addition to the transfer of staff from the military and conservation roles, there is a long history of shared land use for military and conservation purposes. For example, national parks can be important areas for military training. In the UK, Dartmoor National Park has been used for more than two hundred years for training in survival skills and artillery. Kruger National Park was used for military training and as a base for the South African Defence Force during the apartheid era. Militaries have drawn attention to the environmental benefits of putting land under military control, as bases, training grounds, or demilitarized zones in borderlands.[4] The Korean Demilitarized Zone contains a highly biodiverse ecosystem, and the British Army points to the wildlife that flourishes on its military bases around the world.[5] These are important means through which militaries make claims about their positive environmental impact and their commitment to protecting biodiversity.

However, while generating welcome positive stories about the commitment to conservation by militaries, these sites can also serve as a form of "greenwashing," which obscures and masks the destructive practices of militaries elsewhere. As Lunstrum points out, positive coverage relies on a process of erasing history, especially of concealing the evidence of pollution and the associated military activity that produced it. Such stories conceal the fact that environmentally harmful pollutants, which include radioactivity, endure in the longer term.[6]

The engagement between the military and conservation, including the ways that drawing attention to conservation benefits can obscure other environmental harms, is illustrated by the declaration of the world's largest marine protected area in the Chagos Archipelago in 2010. The declaration of the 250,000-square-mile Chagos Marine Protected Area (MPA) in the British Indian Ocean Territory was hailed as a great green news story. The British government presented it as a conservation success that would protect 220 species of coral, 855 species of fish, and 355 species of molluscs. It is a unique environment, and the MPA was deemed to be the best way of protecting the fragile ecosystem.[7] This designation, however, obscured the complex history of the Chagos Archipelago, and it did not address ongoing legal challenges.

The island of Diego Garcia in the Chagos Archipelago is currently a US military base, which was developed after the resident Chagossians were deported by the UK government to the Seychelles, Mauritius, and the UK. The Chagossian people continue their legal challenges to allow them to return to their island. Diego Garcia is a strategically important position for the US. Among other things it is used for refueling US nuclear submarines, it is in missile-striking distance of the Middle East, and it is used by US Airforce bombers operating in the region. When the base itself was built, it involved the destruction of fragile coral reefs, and there are reports of continuing radioactive pollution from the nuclear submarines. In sum, the declaration of the MPA obscured the colonial history of the archipelago, the eviction of the Chagossians and their continuing land claims, as well as the environmental destruction caused by the US military base. It could be argued that those involved in the declaration of the MPA were concerned primarily about important conservation objectives, but recently leaked documents indicate that the protected area was in fact secretly promoted by both the US and UK as the best way to protect the base in the long term, rather than for potential environmental benefits.[8] The case of the Chagos Archipelago reveals the ways that conservation can be part of wider global security initiatives by some of the world's most powerful actors, in this case the US military. This wider context is vital for understanding the reasons for and features of current deployments of militaries in responses to the illegal wildlife trade.

MILITARY TRAINING FOR ANTI-POACHING

The deployments of UK troops on anti-poaching projects in Malawi and Gabon have attracted a great deal of media and public attention and have generated a lot of media interest for the British Army; the governments of Gabon, Malawi, the UK; and the NGOs involved, especially African Parks Foundation in Malawi. It is useful to place these important initiatives in the broader context from which they emerged, as this background helps us understand the two-way exchange of knowledge between conservation and militaries and the wider implications of following this path.

Following high-profile media reports and NGO campaigns about rising rates of poaching of elephants and rhinos, the UK stepped up efforts to tackle the illegal wildlife trade, committing new funding and organizing four high-profile international conferences. One of the flagship areas of support was the deployment of small groups of soldiers from the British Army for ranger training in Gabon and Malawi, with plans to extend to other countries in the future. The genesis of the Gabon and Malawi deployments was quite different. In the case of Gabon, the president of Gabon via Prince Charles asked the British government for support in anti-poaching. The deployment to Malawi came about as a result of the initiative of an army captain, returned from deployment in Afghanistan, who responded to public statements by Prince

Charles about tackling poaching in Africa.[9] The captain wrote to St. James Palace, stating that Prince Harry had flown cover for him during ground operations in Afghanistan, and that he thought he could assist in anti-poaching in Africa via training of rangers with a small team of British soldiers. The palace responded with an offer of funding for his master's research on security and anti-poaching. The former soldier used this support for an intelligence-gathering trip to work out what could be done. With funding from the Prince of Wales Charitable Foundation (PWCF) via a grant to Tusk Trust, the captain engaged in trailling training in Save Valley Conservancy in Zimbabwe, Liwonde National Park, and Majete National Park in Malawi.[10]

A small team from the Yorkshire Regiment went to Malawi in October 2017, but the Malawi deployment had some important differences when compared to the operation in Gabon.[11] First, it was in conjunction with the Malawi Department of Parks and Wildlife and a private entity, the African Parks Foundation,[12] which manages several parks across sub-Saharan Africa by agreements with respective governments. Furthermore, the initiative was supported on the UK side by the Tusk Trust (a UK-based NGO) and by a grant from the UK government's Illegal Wildlife Trade Challenge Fund. Second, the troops provided training on-site in Liwonde National Park, rather than bringing rangers to a central military training facility as was done in Gabon.

The structure of the deployment in Malawi was focused on allowing exchanges of knowledge and skills between equals. The rangers and soldiers were issued the same equipment and rations, and as one trainer stated: "We're essentially putting these guys under the command of that park for the duration that we're there and just embedding them in the normal patrolling that's occurring, just trying to make it more effective."[13]

It is also important to place the deployments in Gabon and Malawi in their wider political context of the time and not see them in a framework of an urgent response on how to respond to the illegal wildlife trade. From 2015, DEFRA funded the deployments of small teams of British soldiers to Gabon and Malawi, with plans to extend to other countries in sub-Saharan Africa in

UK counter-poaching assistance team at the London Conference, 2018. Photo credit: Rosaleen Duffy.

the future. It has become a flagship project for the British Army and for the UK government. For example, Michael Gove, UK secretary of state for the environment, underlined his view that the deployments helped create a sense that the UK was still a world power by stating, "We are funding training missions of this kind around the world, . . . These plans will put the UK front and centre of global efforts to end poaching and the insidious trade in ivory."[14]

However, the operations were also part of a search for a new role for the military in the context of significant cuts in funding and radical restructuring of the UK armed forces. In this sense, the deployments can also be seen as part of the wider International Defence Engagement Strategy, which arose out of the 2015 Strategic Defence and Security Review (SDSR15). It was based on the premise that UK interests and global stability were threatened by growing and complex threats, including terrorism, extremism, cyber security, and the breakdown of rules based international systems. The Defence Engagement Strategy is intended to protect UK interests by projecting UK support and influence internationally to prevent conflict, build stability, and bring prosperity.[15] The deployments for ranger training in Africa can be seen against this background of attempting to develop Britain's global status as a soft power. This is illustrated by the ways that Lt. Colonel Neil Bellamy, commanding officer of 2 RIFLES, described Gabonese poachers as part of cross-border criminal gangs. He sought to draw attention to how the deployment was indicative of the role of the modern army because "this task clearly demonstrates how adaptable the modern Army is and highlights the important role that Defence plays in support of Government Policy."[16]

In August 2016, fourteen members of the Northern Ireland-based 2 RIFLES and a specialist tracker from the Royal Gurkha RIFLES were deployed to Gabon to assist in training sixty rangers and gendarmes operating under the Gabonese Agence Nationale des Parcs Nationaux.[17] The UK troops operated at the military training center in Mokekou, Lope, and rangers were brought to it from the national parks to be trained. However, this arrangement did not work as effectively as hoped because the rangers were being trained in terrain that differed from the protected areas they normally operated in. It was felt that they were being trained only in basic squaddie skills rather than a bespoke, well-tailored package. One of the trainers remarked, "What we found was when we do that, these guys just embed back into a park where 90 percent of the rangers haven't had any training and where they will go back . . . and continue what they were doing [before]."[18]

In contrast, in Malawi, a strategic decision was taken by the British Army team to offer the training in the parks where the rangers already worked.[19] As onc UK civil servant, who was overseeing the deployment suggested, "For Malawi, they were embedding with the teams in their park on their home turf. So, they were working with the terrain and with the kind of equipment they had available and the support they had locally in the park. They have learned from Gabon, applied it to the Malawi project, and had some positive initial results. "[20]

The training in Malawi was explicitly presented as a two-way learning exercise: British troops provide military training to professionalize rangers, and in return rangers help British troops revive their lost tracking and bush skills. In an official statement from the Ministry of Defence, Company Sargent Major Francis Condron said, "Tracking is a fascinating skill, and the rangers are so enthusiastic and happy to share their experience. It is such a useful skill to have, and now having this knowledge I think I'll be able to use it in other situations as a useful survival skill, both with the Army and at home."[21]

There were four key areas in which rangers imparted skills and information to British soldiers: tracking, how to handle encounters with dangerous wildlife, how to sustain oneself in the terrain over long periods, and information on medicinal or poisonous plants. As a trainer remarked, "The guys are constantly learning from the rangers what plants are edible and what things you can supplement your rations with. . . . The rangers are expert in finding water after all, certainly the older guys, and so our guys are learning from them how to find water in a dry environment, certainly during the dry season."[22] These were central survival skills that the army felt it has lost over the previous three decades of international deployments in the Middle East. This was made clear in an article by the defense correspondent for *The Independent* newspaper: "The British Army is building a new team of counter poaching specialists to help allies tackle wildlife crime and try to revive tracking skills lost during the Iraq and Afghanistan campaigns."[23]

Furthermore, the deployments for ranger training have been justified not just as a conservation measure but as a means of responding to global security threats. The need to tackle ivory poaching, and its possible link to funding militias and terrorist groups, resurfaced as part of the justificatory framework. During the first Gabon deployment Julian Brazier, the minister for reserves, stated: "Our Armed Forces are among the best in the world and this important training demonstrates Britain's commitment to Defence Engagement and tackling threats. We have put our full support behind the Gabonese Government in the battle against the scourge of ivory poaching."[24] Minister Brazier's comments directly link ivory poaching as a matter of defense and security, indicating that he regards it as a security threat. This view ties in with the shift toward a growing global narrative around ivory poaching, especially as a global security threat.

The embedded reporting by journalists accompanying British forces has also produced very positive media coverage for anti-poaching training as important for tackling global terrorism; in such coverage the claims about the links between illegal wildlife trade and funding for global terrorism are not subjected to scrutiny. For example, a story by Andy Jones in *The Mirror* (later repeated in *The Independent*) included the claim that anti-poaching in Gabon assists in the fight against jihadists and Boko Haram.[25] The news media reports pointed to the important justificatory narrative for the deployment to UK troops in anti-poaching training; for example, an article in *The Mirror* stated, "Poaching is the new frontier in the war on terror, pitching endangered elephants and the British Army against the forces of evil. . . . Intelligence shows these poachers are linked to extremist group Boko Haram in Nigeria and have killed 25,000 forest elephants in the past 10 years in one Gabon region alone."[26] There is, however, very little evidence for the claims that Boko Haram are engaged in significant levels of ivory poaching (see chapter 4). Furthermore, Boko Haram do not operate in Gabon. This then raises a question about the role of training for anti-poaching in tackling terrorism if those engaged in illegal wildlife hunting in Gabon are not affiliated to Boko Haram.[27] And it carries a risk that the actual poachers and underlying reasons for poaching are missed in strategies that aim to curb Boko Haram.

CONCERNS ABOUT MILITARIZATION

Supporters of the deployments have been careful to address concerns that such training operations can contribute to the growing militarization of conservation. Supporters place their emphasis on professionalization of the ranger units and transfer of skills and knowledge around how to share information, weapons training, and how to seal a poaching site as a crime scene (even though they do not usually engage in crime scene investigations). A public statement released by the Ministry of Defence when the deployment to Gabon was announced noted: "The objective is not to use military training, but instead impart military skills and techniques to the ANPN. . . . This includes navigation skills, patrolling techniques and planning, evidence gathering, first aid, injury prevention in the jungle, weapon handling/safety, fitness and teamwork, live firing, ambush drills and interdiction, and lessons on Gabonese law and Rules of Engagement."[28]

As with the Gabon deployment, those overseeing the operation in Malawi emphasised that it was about professionalization and definitely not about tracking, intercepting, arresting, or shooting actual poachers. Michael Geldard, Britain's defense attaché to several African countries, stated, "We are here to train rangers how to protect themselves from wild animals and not necessarily to shoot poachers."[29] Those closely involved in the deployment on the ground were keen to point out that British Army operatives were issued 7.62mm rifles only as protection from dangerous wildlife, and that they did not have authorization to engage with, or fire at, suspected poachers. There was a keen sense of the political storm that would undoubtedly ensue should a white officer from the UK wound or kill a black poacher in Gabon or Malawi. Therefore, considerable efforts were made to ensure that this would not happen. As one of the trainers remarked to me, "I don't believe MoD has any interest in our guys ever running into poachers. That's not what this is. It's effectively a training course that we can't run centrally so we're delivering in their park."[30]

Similarly, a UK civil servant overseeing the funds for the training operation in Malawi was keen to note, "There are some misconceptions about what it is they are doing . . . they are teaching them [rangers] to get between poachers, or where the poachers might be coming from . . . through interception and tracking techniques."[31] Another pointed out: "I'm aware there is some concern over militarization of rangers and counter poaching, but I don't think this falls in to that category. This is about training them to be safe, cover more ground, think about how they are intervening so they are preventing poaching from happening in the first place rather than catching it as it happens, or trying to pick up the pieces after it has happened."[32]

Media coverage has further cemented a narrative of the modern professional soldier, engaged in professionalization of parks staff rather than in militarized forms of conservation. The media coverage has been primarily in the form of embedded reporting, with reporters invited by the army to observe training operations. For instance, the BBC produced a short film in August 2017 about the deployment by their defense correspondent Jonathan Beale. The film draws clear links between security and poaching. Beale presented the operations from the point of view of the military trainers and the Gabonese government and includes in the film a mock ambush, mock arrests, an interview with President Ali Bongo of Gabon, and an interview with Major Joe Murray of the 2 RIFLES. Major Murray specifically pointed out that the skills developed by the army in previous operations and training events around the world were relevant and transferable to anti-poaching. The news film was accompanied by the headline "Soldiers are sharing tactics from Iraq and Afghanistan to help gamekeepers in Gabon tackle poaching,"[33] a reminder that military-style approaches in conservation have in part been inspired, or even driven, by the transfer of skills from international interventions in the Middle East.

There is an often-unacknowledged politics of race surrounding these training initiatives. For example, in the Ministry of Defence New Team blog on the first deployment in August 2015, the Gabonese staff are referred to as *native* trackers and *native* park rangers (my emphasis) thereby utilizing clear colonial terminology to describe people working in anti-poaching.[34] The images that accompany media coverage have been provided by the MoD and feature a white British soldier in the foreground, watching over a training exercise or pointing out information on a map to black Gabonese park rangers in the background. The "optics" can communicate the notion that the white European soldier-trainer imparts knowledge to black African park guards, rather than it being a two-way process of knowledge exchange.

The narrative surrounding the Malawi operation differed from that of Gabon. MoD statements about the Malawi project explicitly utilize the perspective of a British soldier originally from Malawi, Sargent Kachoka Kingly: "As a local boy, this means a lot to me and my community. This is my home, and to be given the opportunity to come back and protect my country's natural treasures, I feel like a hero. Not only have I been able to support the team with my tracking skills, it has been beneficial to the whole team to have me on board, as a translator and cultural advisor."[35]

In their coverage of the first operation in Malawi, both the *Daily Mail* and *The Independent* relied on a press release from Agence France-Presse, using identical wording: "Under a scorching sun, a team of British soldiers and Malawian rangers sheltered under a tree ready to pounce on their prey: poachers."[36] Looking for poachers as prey was not what the operation was about, but it sounds more exciting than offering training and would be more appealing to the readership. This legitimization of the use of forceful and militarized anti-poaching strategies via the discursive production of the poacher as greedy, criminal, and even evil or subhuman is a good example of a key problem with the current debates about poaching (see chapter 3). Neumann argues that these characterizations of poachers are linked to racial stereotypes, dating back to the colonial era, when African methods of hunting, especially use of traps and snares, were described as cruel and unsporting in order to enclose rights to hunt by "humane and sporting" European hunters using rifles.[37]

Reporting like this also dovetails with the much wider presentation of rangers and their trainers as heroes and can obscure the complex day-to-day roles played by rangers in areas where illegal hunting of wildlife is a significant problem. The idea of rangers as heroes means they have names, stories, and identities, while poachers appear only as a pervasive but anonymous enemy, without names or further detail about who they are.[38] The characterization of hero versus villain obscures the roles rangers can and do play in eviction, exclusion, and perpetration of violence against local communities. It also obscures the ways that rangers themselves may not be in support of military strategies and tactics, and ways they may themselves regard the increasing use of force as counterproductive, leading to an escalation in violence.

Furthermore, within conservation there is a well-known exchange between poachers and ranger roles because they share similar skill sets and often live with or near large wildlife populations. Ex-poachers can be drawn into ranger roles with the promise of formal employment and any additional benefits that go with it, such as housing or education for family members. Rangers can also switch to become poachers or perform dual roles as both poachers and rangers. For example, Lombard shows that rangers in Central African Republic who were provided with military-style training and weapons by external agencies then used them to enhance their political and economic status by joining the wider regional threat economy and engaging in poaching, kidnapping, and looting.[39]

There is an important contextual history to Liwonde Park, which is invisible in the current narrative about the Malawi deployment. This is not the first time that external trainers have worked in Liwonde National Park; a private South African military company provided training there in the 1990s. However, in just two years (1998–2000) the parks staff in Malawi who had received the training, were implicated in 300 murders, 325 disappearances, 250 rapes, and numerous instances of torture and intimidation.[40] No mention is made of this history in statements and media coverage about the current deployment. It offers an important lesson—that training does not necessarily lead to professionalization of rangers, and that any training does not operate in a vacuum. And it intersects with, and possibly deepens, existing dynamics. Furthermore, it is clear that the military trainers are deployed to areas where there is no elephant or rhino poaching, such as Majete National Park in Malawi,[41] and thus their anti-poaching operations are often focused on those engaged in extra-legal subsistence hunting and fishing. Supporters argue that the British Army training is different from that of a private military company, and that it may be held to higher standards. This misses an important point, that any new training of resources does not necessarily produce a shift, but instead interacts with the existing context. Further training and provision of equipment, including arms, may simply produce a group that is more effective at harming local populations and wildlife in the area.

PRIVATE SECTOR SECURITY ENGAGEMENTS

The growing engagement between militaries and conservation extends well beyond the deployment of standing armies of states. Conservation mirrors wider developments in the security sector. More broadly, the expansion of PMCs offers a range of services from bodyguards to humanitarian assistance to protection of oil platforms.[42] High-profile examples include Erik Prince's Blackwater, later called Academi, but there are many others. The privatization of security and contracting out roles that are normally fulfilled by national armies became much more common and normalized following US-led interventions in Iraq and Afghanistan. This wider dynamic of the privatization of security has also shaped conservation practice as these new companies search for new markets for their skills. This search was then met with apparent *need* for security services in conservation, as conservationists expressed concerns that wildlife crime, poaching, and insecurity constituted key and rising threats to some iconic species.

The range of organizations operating in conservation has expanded, bringing new approaches and practices. Several organizations have provided new roles in conservation for veterans from wars in Afghanistan and Iraq.[43] These include US-based VETPAW, which trains ex-military to work in anti-poaching. Another good example is International Anti Poaching Foundation (IAPF), set up by Damien Mander, who originally trained in the Australian Navy and the Army Special Forces; he later worked in Iraq with PMCs.[44] When he set up IAPF he stated that he would use it to apply the lessons he learned in Iraq to anti-poaching first in Mozambique and then in Zimbabwe.[45] Following his initial work, Damien Mander set up an all-women anti-poaching unit called Akashinga in the Zambezi Valley in Zimbabwe. It has received a great deal of media attention, including an eponymous *National Geographic* documentary.[46] These wider developments are worth further examination.

It is useful to reflect on what motivates people to move into conservation following a military career. As with Lombard and Tubiana's reflections on game guards,[47] the shift from the military into conservation can be seen as part of the wider career path for some individuals. Once they have left the military, they have several options, including taking more mainstream and lucrative contracts for big name PMCs. In fact, one private contractor engaged in training US military veterans in conservation pointed out that they do have commitments to wildlife, and that they could easily earn far more in private military sector: "On average, our veterans, the guys and girls that go out on the ground and do this work, they can go and be working for Blackwater or Triple Canopy. . . . They could go and be working for them in Afghanistan and making $500 or $600 a day, not counting per diem. A lot of our guys and girls on the ground, they volunteer their time. They're not even getting paid. Some of them are making—if they're lucky—they're making that per week. I think some of that is important. It's not that it undervalues what they're doing. A couple of things come into play with that. The threat level is not as high as it is in, say, some of those other conflict areas, that's first and foremost. Secondarily it's . . . about their position and the reason that they're there."[48]

Many of the ex-military personnel engaged in conservation made the shift because of a genuine commitment to wildlife and because they feel their skills, including a "can-do" attitude, can be of use. One war veteran who now works in conservation, explained how his initial motivation came about: he had watched a documentary on CNN which showed the (now-infamous) footage of an elephant with its trunk severed and a rhino in South Africa, which was still alive and bleeding profusely following the removal of its horn by poachers. He described "all of my emotions from war that I had stuck into a jar, screwed on as tight as I could, super-glued it shut, that rhino, that elephant, and that whole documentary, unscrewed that jar."[49] He recognized, however, that feeling such an intense response, a need to do something to save these animals, was not necessarily the best motivation, and that it had led to some mistakes and challenges for his organization when he first set it up. He emphasized that several years on, his organization does not accept trainees who describe their motivations in this way because they still need to address the deep emotional effects of their experiences of warfare.

Furthermore, recruits can be attracted to conservation because the work can have positive impacts on mental health. Indeed, VETPAW explicitly states that its work can help veterans come to terms with PTSD that can follow combat experience. The director of one of these organizations, a US veteran who served in Iraq, was proud that conservation work could have beneficial mental health effects for staff; he remarked, "It's really beautiful, they are saving the animals, but the animals are also saving them."[50] The healing and regenerative effects of being in an African context were also cited as having a positive impact. As one trainer commented: "We don't bring veterans out that are just shattered by PTSD. PTSD is something you always live with, you manage it, you cope. We have veterans that come out and they're able to decompartmentalize and make sense of everything that they've gone through because, you're in Africa, it's healing."[51]

This is important, because, as White suggests in his study of veterans working in private sector security, PTSD presents a challenge for the private sector. On one hand, private operators can regard treating PTSD among their contractors as a "cost" that needs to be factored into their business model, but, on the other, it is a wider political or social issue produced by specific civil-military relations. There can be a wider commitment to dealing with PTSD as a result of combat operations carried out as part of a national army, yet there is a real challenge around whether private companies take responsibility for caring for and paying for treatment for staff who display PTSD while they are under contract and, indeed, afterwards.[52]

The engagement of private contractors goes beyond simply using veterans for specific operations or skills or providing them with employment opportunities and a chance to address PTSD. These veterans bring and encourage particular ways of thinking about conservation, which are more oriented toward their own worldviews and may not match those of local communities and national governments.[53] This includes categorizing types of people or groups as enemies as well as areas or "things" to be defended.

A security-oriented approach also means that ranger training can be transformed away from the usual roles of park rangers and toward a focus on military-style training and counterinsurgency tactics.[54] Such a shift inevitably changes the nature of potential recruits, attracting applicants who might be more amendable to military-style approaches and are less interested in the ecology of protected areas. Indeed, one private contractor made it clear that there was a genuine interest amongst rangers in learning more about the techniques of US warfare. He commented: "I found that a lot of park rangers, and a lot of Africans in general, were very up to date and up to speed with the American wars. They were blown away that a US veteran was coming over to share some of these skills with them."[55]

In the longer term this fundamental shift toward recruiting those interested in enforcement and military strategies could risk the broader ecological health and integrity of protected areas. Indeed, the expansion of private parks, and engagement of NGOs and private companies to run protected areas, facilitates and extends this private sector approach to security. Private reserves, especially in South Africa and in Mozambique, commonly contract private security companies to secure their reserves and the assets (wildlife, buildings, vehicles, and staff) within them.[56]

The proliferation of private sector conservation security companies points us to an underacknowledged theme: the political economy of privatized security in conservation. Despite the scandals surrounding the use of private intelligence and security operations in the past, such as KAS Enterprises and in Liwonde, the practice is expanding.[57] The question of who profits (financially) from the engagement of private security contractors in conservation is often overlooked. Massé's and Lunstrum's study of anti-poaching strategies in and around Kruger National Park

Ranger (left) from an anti-poaching NGO patrolling with a member (right) of Mozambique's Environmental Police, formed in 2014. Photo credit: Francis Massé.

demonstrates how such approaches to conservation open opportunities for new sources of profit for the private sector, which they call "accumulation by securitization."[58] The neoliberal conservation phase paved the way for this normalization and acceptance of private sector security. It has shaped the direction of conservation on the ground and has the potential to constrain other options in the future.

Further, there is an underacknowledged politics of race and of gender in many of these new initiatives. All-women anti-poaching units, such as the Black Mambas in South Africa[59] and Akashinga in Zimbabwe, are presented as a means of empowering women; it is a story that captures the attention and imagination of external audiences with the argument that it creates a win-win, both for conservation and for women, by helping them escape domestic violence and giving them opportunities to earn an income and gain ranger training—but from male military trainers.[60] Further, there is a politics of race which is generally not discussed, but which is present in the visual coverage of Akashinga: white male military trainers imparting knowledge to African "recipients," which draws on the long history of the idea of a white saviour bringing aid and support to needy Africans.[61]

The expansion of private military companies operating in conservation, especially in sub-Saharan Africa, has generated concerns about the variation in practices, professionalism, and oversight. Despite the proliferation of these organizations there is no common code of practice or set of guidelines against which they can be benchmarked. Again, much could be learned from the humanitarian sector, especially since it also often works with military actors in areas of armed conflict, operates under intense pressure, and faces threats from armed groups. For example, the International Committee of the Red Cross (ICRC) has drawn up guidelines for those engaged in security and humanitarian protection work in areas of armed conflict.[62] There are no guidelines for those offering protection work or military-style training in conservation, but such guidelines could provide a very useful benchmark.

The variability in the quality of groups operating in conservation mean that some have sought to distance themselves from other companies or individuals in order to protect their reputations. One representative of a conservation organization that trains and deploys veterans expressed concerns that so many NGOs had been established by US war veterans, with little regulation and oversight from either the US government or the governments of the countries in which they operate. The variability in quality of work and types of practices, he said, were giving all such organizations "a bad name." In his organization, recruits are carefully vetted and undergo a rigorous and intensive three-week mental and physical training program. Only a very small proportion of applicants are approved to work on the ground in African countries, and when they do so they are deployed with the full knowledge and approval of the relevant government.[63] This has raised concerns amongst PMCs that their roles may be more generally misunderstood, as one private contractor commented, "There was a misconception, is a misconception, that organizations like ours are there as mercenaries."[64] Another remarked, "We don't want to be painted with that same brushstroke because, as you can see, it happened. These other organizations can taint a reputation that other people have worked so hard to build. . . . Guy goes on YouTube and watches all these military videos and they think they're Ricky Recon, GI Joe, and they're going to go make their own YouTube videos and train rangers and this and that. . . . It is so important that the conservation world gets a handle on who exactly is coming out. There needs to be more oversight."[65]

The expansion of private sector military and security operators in conservation has accelerated as a result of the convergence of factors: the availability of demobilized military personnel from interventions in the Middle East, the wider normalization of the use of private military companies, and the broader privatization of security, coupled with the security turn as a result of a sense of urgency needed to respond to rising rates of poaching and trafficking. Using a political ecology lens, these shifts are interesting because they are slowly changing the nature and practices of conservation. They also raise wider questions, including what are the implications for state sovereignty of giving control of parts of their territory to armed private sector actors? Under whose authority might PMCs use deadly force in pursuit of conservation? Are employees of such companies to be indemnified, as rangers often are, or not? What are the responsibilities of the companies toward their employees if they experience further mental or physical health problems? To date, the lessons from the much longer standing use of private security companies in delivering humanitarian assistance and support, or in post-conflict peace-building, have not featured prominently in discussions about the wider implications of these shifts.

THE NONHUMAN IN SECURITY AND CONSERVATION

The issues presented in this book, and indeed much conservation debate, revolve around how humans are conserving wildlife. Yet we must not overlook the important work undertaken by trained animals; they have been used in conservation and security, in particular in defense of other animals. The use of working animals builds on a long history: one only need think of the dogs, horses, camels, and elephants used in warfare or the role of carrier pigeons for communications and dolphins for marine patrols. Dogs and African pouched rats are particularly important participants in security-style approaches to the illegal wildlife trade. For example, US FWS provided US$100,000 for a trial project to train pouched rats, animals that have been able to sniff out explosives, to identify illegal wildlife products at ports in Tanzania.[66] The development of animals being trained as part of strategies to tackle the illegal wildlife trade is fundamentally restructuring relations between animals and relations between animals and humans. So what are the implications for nonhuman nature and how we might understand shifting human-animal relations?

Dogs are used in anti-poaching operations and as sniffer dogs. Each task requires different kinds of dogs with distinct kinds of training. A 2015 assessment of the use of dogs in anti-poaching in Africa indicated that the last five years has seen a rapid increase in canine law enforcement programs in Africa and that, overall, these programs have had mixed success.[67] Sniffer dogs have been identified as important by organizations engaged in tackling the illegal wildlife trade.[68] There is a long history of using trained detection dogs for searching for explosives and drugs by law enforcement agencies, militaries, and customs authorities. For example, the US-based Working Dogs for Conservation (WD4C) searches in shelters and rescue centers for appropriate dogs to train. High-energy, intelligent dogs make good candidates for sniffer dog training, but these qualities make them very challenging as domestic pets.[69] WD4C has provided five detection dogs to seek out illegal wildlife products and weapons in Luangwa Valley. In addition, it spent several months training handlers from Conservation South Luangwa and the Zambian Department of National Parks and Wildlife.[70] The dogs are sourced from US shelters, rather than locally, despite the additional costs of training and transporting them to Africa, because "village dogs simply don't have the drive to do this kind of work. There are only a handful of suitable and reputable kennels in Africa. Most are focused on selling security and military dogs, so they're not as well socialized as a conservation dog needs to be. Plus, they generally sell those dogs for much more than what it would cost us to source a dog in the US."[71] WD4C maintains that it is saving wildlife by saving dogs that would otherwise be euthanized in the US because they are unsuitable domestic pets, and the dogs "do it for the love of the ball."

This raises an interesting set of themes around the training and use of animals to conserve or *save* other animals. It infers that the domestic working animals choose or enjoy their work. In the arena of detection dogs it could be a persuasive argument, but it is less convincing when considering the use of dogs in direct anti-poaching operations. This kind of training and use results in more negative dynamics in human-animal relationships: the production of dogs for profit making in the private security industry, the welfare of the dogs themselves, and the impact on the dog handlers and trainers of working closely with dogs trained to attack.

These dynamics are illustrated by the case of Killer, the anti-poaching dog hero. Killer is what is colloquially referred to as a "bite dog," one that is trained to attack a person on the command of its handler, biting and pinning down a suspect until rangers can secure and arrest the person.[72] In 2016 Killer was awarded the People's Dispensary for Sick Animals (PDSA) Dickin Gold Medal for his work in anti-poaching in Kruger National Park, where he had intercepted 115 poachers.[73] The Dickin Medal is awarded to animals for gallantry and actors of bravery during military conflicts.[74]

Media interest in Killer obscured the political economy of Killer's existence. Killer was bred and trained for sixteen months by the Paramount Anti-Poaching and K9 Academy, based outside Johannesburg, and sent to Kruger National Park when he was eighteen months old. Another dog in his training cohort was Arrow. That dog entered the *Guinness World Records* as the first dog to do a parachute jump.[75] Both Arrow and Killer have been with their handlers since they were puppies in order to establish the strong bond needed for the dog to follow commands.[76] Paramount is South Africa's largest private defense and security company, and it has several contracts to provide anti-poaching support in the park. It designs, develops, manufactures, and sells military vehicles, surveillance systems, and weapons systems around the world.[77] It is owned by the Ichikowitz family, and dog training is listed under its philanthropic arm, the Ichikowitz Foundation, as part of Paramount Group's giving back strategy.[78] As Lustrum suggests, such strategies are central to the ways that defense contractors expand into new markets and as a means of greenwashing the fact that they are fundamentally weapons manufacturers.[79]

Little information is available on how the dogs are housed and cared for. Parker's review of dogs used in conservation in Africa found there were no accepted standards for handler training, dog training, or dog care, and no best practices have been established for the field. Many programs work in isolation from one another, without the benefit of ongoing training, collaboration with other programs, or knowledge of what has led to success or failure for other canine programs.[80] Parker's review team received reports of dogs that were trained only for several weeks, as opposed to the months of training recommended, before being delivered to poorly prepared handlers. These teams failed almost immediately, since handlers require as much, if not more, training than their dogs.[81] One overlooked challenge highlighted by the review is that certain breeds are associated with colonial histories in Africa and are chosen precisely because they are intended to be intimidating to African communities. The integration of conservation and security has facilitated this kind of reshaping of animal lives to make them compatible with the needs of the security sector, which is restructuring relationships between humans and animals, and between trained animals and wildlife.

The current engagements between the military and conservation are built on long-standing historical linkages. The skills of rangers and the military do overlap somewhat, which is why there was a degree of interchange between the two in the colonial period in places like Kenya and Tanzania. It is important to examine the practices and implications of the growing engagements between the military and conservation. These include the deployment of national armies for training and professionalization, as in the case of the British military in Gabon and Malawi. These deployments need to be seen in a wider context of a search for a new role for militaries, including the need to learn (or relearn) skills in readiness for future conflicts. The two-way knowledge exchange between soldiers and rangers indicates how conservation can be an important means through which nature becomes entrained to work for global security concerns. There is an important political economy surrounding this which is often overlooked. It is apparent in the development of thriving private sector security companies and a new breed of conservation NGOs staffed by former military. There are several examples of companies staffed by war veterans from international interventions in the Middle East, offering their skills to conservation. It is important to place this fact in a wider process of the privatization of security around the world.

The integration of conservation and security is also restructuring relationships between animals, humans, and the environment. While there is a long history of using animals in warfare and in the security sector, in this current phase they can become part of a wider political economy of greenwashing for weapons manufacturers, thereby contributing to the wider militarization of society. This, finally, brings us back to one of the core arguments of this book, that the integration of security-oriented approaches in tackling the illegal wildlife trade is fundamentally reshaping conservation in ways that do not necessarily tackle the underlying drivers of trade.

EPILOGUE

A Way Forward

Security is now a defining feature of conservation. The sense of crisis generated by illegal wildlife trade is a perfect illustration. There is a fear among conservationists that rising rates of poaching and trafficking will drive some of the world's most iconic species to extinction. This fear rapidly galvanized support for clear and decisive action to save them, a response now magnified by COVID-19, with conservationists concerned that the economic implications of the global pandemic will encourage poaching, reduce revenues to conservation NGOs, and place biodiversity conservation lower on the list of national and global priorities. There is no doubt that there is a need to tackle the illegal wildlife trade, but the central question remains, how? The answer that emerged in the wake of the poaching crisis of the late 2000s was faith in security-oriented approaches. The illegal wildlife trade was redefined and presented as a threat to security and stability, one that could be tackled via methods normally used for dealing with organized crime networks and armed groups. These include counterinsurgency, intelligence gathering, technological solutions, militarization, anti-money laundering, and enhanced law enforcement. All this was ushered in, supported by donors, NGOs, philanthropists, and businesses. However, more critical reflection is needed.

Much effort and energy has understandably gone into gathering information on the dynamics and patterns of the illegal wildlife trade: source countries, trafficking routes, consumer demand, seizure rates, and impacts on wildlife populations. All of this is vitally important work, and it helps build up a bigger global picture of the illegal wildlife trade. However, little attention has been paid to understanding the patterns, dynamics, and implications of the policy responses to the trade. The sense of urgency about the impacts of the illegal wildlife trade on elephants, rhinos, tigers, and, more recently, pangolins forged an alliance between conservation and security. This has had substantial implications for conservation more widely, for wildlife, and for people, and so it requires careful scrutiny. In this book I have investigated these dynamics more closely, to offer critical reflection, point to the potential pitfalls, and document some of the injustices they produce.

First, a political ecology lens helps us make sense of these shifts in conservation and highlights their wider implications. I used this intellectual foundation to focus on excavating and understanding the ideas and values behind current strategies, and to indicate how global security can be underpinned by our relationships with the environment. Second, I have analyzed the range of actors and power dynamics that produced the growing integration of security and conservation, allowing an understanding of the politics surrounding how, and why, illegal wildlife trade is framed as a security threat in ways that have changed conservation thinking and practice. Third, I have highlighted the injustices produced by this security turn, especially for less powerful actors. The more security-oriented approaches in conservation have been explained and justified as a last resort response to an emergency situation. In many ways conservationists, especially those operating in contexts of protracted armed conflict or facing armed poaching groups, feel they have no other option but to defend themselves by partnering with militaries, using intelligence networks, relying on new technologies, or in some cases using more forceful and militarized responses. Essentially this book has explored and examined the challenges that such a security turn produces, especially for the longer term. It provides an opportunity to pause and reflect on the outcomes produced by that sense of urgency.

This book cast light on a range of hidden dynamics of security-oriented responses and what they mean for conservation, people, and wildlife, in the longer term. The aims of security and of conservation can be very different. Conservation aims to conserve species and habitats, as well as to prevent extinctions as a result of unsustainable and illegal trade in wildlife. Traditional approaches to security are about preventing or responding to the threats to states. It is not at all obvious that these two sets of divergent logics should be fused together in responding to the illegal wildlife trade. But they have been fused together, via redefining illegal wildlife trade as a serious crime and a security threat. The main argument in support of this thinking is that illegal wildlife trade is conducted by criminal networks and/or generates threat finance for armed groups.

It is essential to analyze these issues because of the sheer scale of changes in conservation as a result of its growing engagement with security. The integration with security renders visible three sets of dynamics: first, how it is changing what conservation can be and what it cannot be in the future; second, how it is fundamentally restructuring human relations with the environment, and relations between wildlife and the environments they inhabit; and third, how taking a security approach to the illegal wildlife trade, ultimately, will not solve the problem of species losses.

The first theme in the integration of conservation and security via strategies to tackle the illegal wildlife trade, is changing conservation practice on the ground. It is defining what conservation *can* be and, more importantly, what conservation *cannot* be in the future. The new security-oriented approach has opened an opportunity for new actors to enter into conservation, thereby reshaping it in the longer term. There are new partnerships between conservation and private military companies, weapons manufacturers, intelligence services, and tech companies. These kinds of connections can cut out or alienate other potential collaborators, including communities living near protected areas, environmental activists, and grassroots development NGOs.

Further, it has reenergized and embedded militarization in conservation. Conservation has a long history of engagement with militaries and militarized approaches, but the growing use of forceful methods has significant implications for people and wildlife. The impacts for people can be twofold: for conservation professionals, it changes their daily work, attracts in different kinds of applicants for ranger roles, can lead to higher rates of workplace stress and even PTSD; for people at the receiving end of militarization, it can mean human rights violations, including murders, rape, torture, beatings, harassment, eviction, and dispossession. Turning a blind eye to such outcomes on the ground is unacceptable. They are often explained away and minimized as the result of poor practice by a few rotten apples, but they are part and parcel of militarization as an approach. For wildlife, militarization can mean a restructuring of the relationship between wildlife and human communities. For example, rhinos are sometimes allocated armed guards that vigilantly watch over them. Elephants can be made subject to constant remote monitoring, and birdlife can be disturbed by drone surveillance. This is supported and maintained by the wider political economy of the security industry. Increasing levels of funding available from donors, NGOs, governments, and the private sector have facilitated the entry of new security-oriented actors into conservation. The availability of new streams of funding has encouraged the development of new markets for security-related products and services. Security-style approaches can result in supporting and maintaining the position of social, political, and economic elites, who then shape and implement militarized responses. Even the alternative livelihoods- and development-focused approach of community-based natural resource management can be integrated with the security logics, producing community-based militarized conservation. Meanwhile the voices of marginalized peoples living with wildlife are rendered silent, while militarization reenergizes an old and deeply problematic fortress conservation model.

Furthermore, militarized forms can be the means by which state agencies extend authority and control over resistant and unruly areas and populations. The war for biodiversity is then transformed into a new phase, a war by conservation. In this war, conservation becomes a willing partner and ally in prosecuting a diffuse, endless US strategy of the War on Terror. This is anchored in contemporary anxieties about global security threats, yet there is limited evidence (in the public domain) of the use of the illegal wildlife trade by terrorist networks, and even less discussion of what the term "terrorist" actually means in this context. As it stands, the links are assumed as evidently or intuitively correct by proponents, leading to strategies that can be targeted at the wrong groups, the wrong activities, and the wrong places. In seeking to use conservation to tackle terrorist threats several armed groups are invoked as threats to security and wildlife, for example, Boko Haram, Janjaweed, Al-Shabaab, and the Lord's Resistance Army. The deployment of British Army counter-poaching teams to tackle ivory poaching and Boko Haram in Gabon is a good example. It involves more than just a transfer of knowledge and skills from the military to conservation professionals; it is about militaries (re)learning skills and seeking new roles on the context of restructuring of armed forces following austerity cuts and western interventions in the Middle East.

Militarization is only one aspect, albeit high profile and spectacular, of how security-oriented approaches are changing conservation. There are many other ways in which security logics are creeping in and expanding through conservation, including mundane, everyday practices. Intelligence-led approaches are part of the wider integration of conservation and security, which also include the use of informant networks and counterinsurgency techniques. It is important to place these approaches in a wider context and outline what the possible challenges and pitfalls might be. One of the most significant challenges is the potential breaches in confidentiality and anonymity for people providing sensitive information to NGOs interested in gathering highly sensitive information. Such breaches, and ineffective handling of sensitive data, put informants and conservation staff at enhanced risk of reprisals.

The security-oriented approach to conservation has encouraged a rush toward technological solutions that rely on new forms of mobile and surveillance technologies. These include camera traps, remote-controlled cameras, drones, remote sensing systems, and surveillance of mobile phone communication. Some of these technologies, such as camera traps, are more widely adopted for monitoring wildlife populations and can be significant contributors to understanding the movements and status of animals. But technologies also produce new ways for conservation authorities, NGOs, and philanthropists to monitor the movements and activities of rangers from afar and as such can create a twenty-four-hour working environment that means staff are always "on call," with the additional stress that produces. The use of these technologies has facilitated the entry of new private sector actors, such as security and intelligence companies, into conservation. These actors bring security-oriented modes of thinking with them. The faith in technology provides more space for the private sector to enter in to conservation practice and to profit from it in financial and reputational terms. Therefore, it builds on the earlier neoliberal phase of conservation, which is very apparent in the involvement of philanthropists from the tech sector. It is important to learn lessons from overreliance on big data, algorithms, and AI found in other sectors. Algorithms are developed out of a complexity of human decisions, which means they are not neutral instruments but are shaped by wider power structures and can be used to reinforce or deepen existing inequalities.

The second theme that a political ecology lens reveals is how the convergence of conservation and security is fundamentally *restructuring* human relations with the environment and relations between wildlife and the environments they inhabit. Recognizing this is important because we need to grasp how framing conservation as a security issue drives decisions that have material consequences for wildlife and people. The imperatives of security and of conservation do, in effect, lead in two different directions. However, security imperatives tend to win out over conservation concerns. As a result, responses to the illegal wildlife trade can become part of wider security strategies at national and international scales. This is facilitated by the definition of illegal wildlife trade as a wildlife crime, a serious crime. For example, it informs who or what can be perpetrators of wildlife crime, and it helps produce and embed the idea that the illegal wildlife trade is perpetrated by bad guys, criminals, armed groups, and even terrorists. This thinking makes relying on forceful responses to poaching and trafficking more palatable, even if greater use of violence is aimed at the wrong actors for the wrong reasons.

Settling on this approach is problematic because it distracts from the underlying drivers of the illegal wildlife trade. It also diverts attention and resources away from alternative ways of designing effective conservation responses. If conservationists started from the idea that unsustainable illegal trade is driven by dynamics of global inequalities, then effective interventions would revolve around tackling an unequal economic system, providing sustainable livelihoods, opening genuine opportunities for people, and pressing for the wider structural changes needed at multiple scales. This would be a form of conservation that takes social justice seriously and uses it as a starting point, rather than as a useful "add on" once the animals have been saved. If we start from the idea that illegal wildlife trade is driven by unsustainable demand because of growing wealth and inequality, and because of historical cultural attachment, then of course effective interventions might revolve around tackling inequality and persuading consumers to switch away from using wildlife products. This is precisely why framings matter. If the illegal wildlife trade is the result of inequality and an environmentally unsustainable and unjust global economic system, then it prompts one set of policy responses. If it is regarded as a serious wildlife crime and a security threat, then it produces a very different policy response. Responses that focus on treating the symptoms (poaching and trafficking), rather than the underlying drivers, will only ever have limited success. It may slow the rate of decline, or even stop a few local extinctions, but a security response will not change the overall pattern of global species declines—because they are the result of much wider structural processes. Further, a security-oriented response may even contribute to supporting and strengthening those problematic structures.

The third theme is that taking a security approach to the illegal wildlife trade, ultimately, will not solve the problem of species losses. This is not to say that security-oriented approaches have no positive impacts; such an argument would be naïve and inaccurate. Security approaches will inevitably have uneven outcomes, differing by place and context. A focus on numbers of animals, sizes and frequencies of seizures by law enforcement agencies, numbers of arrests, or reductions in rates of poaching, while important, only provide a partial picture of the success of this approach. Ultimately, it is crucial that any claims to success also encompass the ripple effects of such strategies, especially if they produce social injustices on the ground. For instance, can we consider it a success story if a violent, forceful approach significantly reduced elephant poaching in a protected area? Many conservationists will see it as a resounding success because the indicator of success is a reduction in poaching and an increase in numbers of elephants. But what if that same approach alienated local communities and resulted in human rights abuses at the hands of conservation professionals? It is not a success if we also include indicators of human rights or social justice. It is vital that when deciding what constitutes a conservation success, we take into account the significant, negative impact on people—and, more importantly, include conservation strategies that take the interests of animals and people seriously.

Finally, on a more hopeful and positive note, the security-oriented approach to tackling illegal wildlife trade is relatively new, it has been built and embedded over the last ten years, and just as it was built, it can also be dismantled and replaced with something else. The question remains: what would replace it? What would be more effective in tackling the illegal wildlife trade? It is certain that conservation, and its supporters, need to rethink their responses. I do not claim to offer the definitive answers, nor should I, but there are some key themes that will be important for generating a new way forward for conservation.

THE WAY FORWARD?

The global COVID-19 pandemic has produced a fresh sense of urgency about tackling the illegal wildlife trade. Current evidence points to transmission from wildlife to humans as the most likely source of COVID-19, and as a result there have been high-profile calls from conservation and animal welfare organizations for total bans on wildlife markets to prevent future pandemics. What is clear is that relations between humans and nonhuman animals need to be rethought, not just in terms of whether wild animals are traded and consumed but also in wider systems, such as intensive farming. There is a need for a fundamental rethinking of how we approach conservation, anchored in more integrated understandings of human relationships with the nonhuman sphere. There are three key themes that should inform the development of fresh approaches to wildlife conservation.

First, it is vital to decolonize conservation. The answers for future models of conservation are unlikely to come from the established conservation sector, steeped as it is in colonial histories that still shape many modes of its thinking and practice. The history of separating humans and nature into discreet parcels of "wilderness" was produced by dispossession, evictions, and ongoing displacement, often accompanied by considerable degrees of violence. In decolonizing conservation there is an urgent need to recognize and offer redress for this colonial history. This goes beyond offering participation, giving voice to, or including those dispossessed as just one stakeholder among many; it means material forms of redress for past and ongoing injustices. Further, it is vital that indigenous communities take the lead; these communities have often been dispossessed in the past or currently face threats to their land rights, sometimes from conservation initiatives (as in the case of BaYaka communities in Congo). It is also necessary to acknowledge, pay attention to, and address the often-hidden politics of race in conservation. When we think of "conservation heroes," the list of household names is likely to be predominantly white and from the global North: Richard Leakey, Dian Fossey, Jane Goodall, Joy Adams, David Attenborough, Emmanuel de Merode, to name just a few. Celebrity naturalists and "explorers" who grace our television and movie screens, our magazines and newspapers, our social media are also most likely to fall in this category. Films like *Virunga* too often place a white savior in the center of the conservation story. In academia, conservation research reflects the structural politics of race, gender, and sexuality; embarking on research in particular places can rely on social networks structured by race and social position, as Ogada shows so well in his book *The Big Conservation Lie.*

Second, conservation needs to tackle the underlying drivers of biodiversity loss, especially the global economy, much more squarely than it currently does. In many conservation circles there remains a commitment to the idea that population growth is "the problem." Indeed, some of the conservation superheroes cited above (Attenborough and Goodall) are part of Population Matters, which focuses on reducing population as a means of achieving sustainability.[1] This approach fails to grapple with the dynamics of a global economy that is focused on growth and consumption, identified by the IPBES Assessment as a driver of biodiversity losses.[2]

Third, there is an urgent need to find new ways of living with and thinking about nature, rather than seeing it as a series of services to meet human needs. Doing this will require a fundamental rethink of humanity's relationship with nonhuman nature. There are several debates that link in to this. One of the most high-profile is E. O. Wilson's Half Earth Project,[3] which aims to place half the global surface under some form of protection in order to ensure that 85 percent of species can survive. Similarly, there have been recent calls for "30 by 30," that is, protecting 30 percent of the planet by 2030. However, both models replicate the problematic separation of humans and nonhuman nature, which has led to the criticism that "Half Earth" should be replaced by a "Whole Earth" approach, or "Convivial Conservation," which are much more integrated approaches.[4] Ashish Kothari has developed the idea of eco-swaraj to encapsulate resistances to the dominant model of development.[5] Others have called for compassionate conservation, which brings animal welfare and conservation into conversation with one another.[6]

Going into the future, and designing and supporting conservation for a sustainable world, the answers are unlikely to come from the offices of global conservation NGOs or academics based in the global North. Instead we must look much more widely and support the voices of those who have historically been pushed out and marginalized by conservation.

Finally, the security turn that came from the integration of conservation and security in responses to the illegal wildlife trade matters precisely because such modes of thinking take conservation down a particular route, from which it will be difficult to return. In the high-pitched debates about the need to act urgently to save species threatened by the illegal wildlife trade there has been limited critical reflection on the responses to it. Little attention has been paid to understanding how and why conservation has taken a turn toward security as a solution to illegal wildlife trade. This book attempts to address this shortcoming. The focus on security, organized crime, and terrorism can distract attention and divert resources from tackling the root causes of losses that are driven by a highly unequal global system. Wealth can drive demand, while poorer communities will often end up as the suppliers at the other end of the chain. Poorer communities take the risks, receive a fraction of the value, and then live with the more hostile and forceful approaches of those organizations and institutions that seek to protect wildlife. Wrapping conservation in a security framework has produced a rationale for wider forms of surveillance and monitoring of people, animals, and ecosystems at one end of the spectrum. However, at the other end of the spectrum, it facilitates and encourages violent enforcement strategies, which can lead to exclusion, dispossession, human rights abuses, and the use of deadly force. As this approach continues to develop, we need to ask ourselves if this is the sort of conservation we want, even when it is incapable of addressing the fundamental drivers of illegal wildlife trade.

NOTES

Interviews are coded to provide anonymity to interviewees. The serial number refers to the date of interview, Work Package (or section) of research for the BIOSEC project, and interviewee number.

PREFACE

1. The sculpture was supposed to represent humans living in harmony with wildlife; but from the side it looked like a man holding an erect penis, which caused a great deal of laughter, especially among women at the conference.

2. Massé et al., "Conservation and Crime Convergence?"

3. *The Guardian,* "White Saviour," accessed 25 March 2019, https://www.theguardian.com/global-development/2018/oct/12/prince-william-accused-of-white-saviour-mentality-over-wildlife-conservation-video-tanzania.

4. Brockington, Duffy, and Igoe, "*Nature Unbound.*" Brockington and Duffy, "Capitalism and Conservation."

5. Ashaba, "Historical Roots." Day, "Sister Forces."

6. See, for example, Sommerville, "Ivory"; Rademeyer, "Killing for Profit"; Nuwer, "Poached."

CHAPTER 1. CONSERVATION AND SECURITY CONVERGE

1. Game Rangers Association of Africa, "Media statement," accessed 8 May 2017, http://www.gameranger.org/news-views/media-releases/170-media-statement-the-use-of-military-and-security-personnel-and-tactics-in-the-training-of-africa-s-rangers.html.

2. Interview with 2017_07_18 WP 1.16c.

3. Wyatt, van Uhm, and Nurse, "Differentiating Criminal Networks."

4. Büscher and Ramutsindela. "Green Violence." Marijnen and Verweijen, "Selling Green Militarization." Lombard. "Threat Economies."

5. Mogomotsi and Madigelele, "Live by the Gun, Die by the Gun."

6. McClanahan and Wall, "Do Some Anti-Ppoaching." Marijnen and Verweijen, "Selling Green Militarization."

7. https://news.mongabay.com/2017/10/attacks-on-militarized-conservation-are-naive-commentary/, accessed on 18.04.20. Mogomotsi and Madigelele, "Live by the Gun, Die by the Gun."

8. IPBES, *Global Assessment Report on Biodiversity and Ecosystem Services,* 2019.

9. O'Brien and Barnett, "Global Environmental Change." Suhrke, "Human Security." Commission on Human Security, *Human Security Now.*

10. O'Brien and Barnett, "Global Environmental Change and Human Security"; Amartya Sen, *Development as Freedom* (Oxford: Oxford University Press, 1999). Kashwan, *Democracy in the Woods.* Duffy et al., "Links between Poverty and Illegal Wildlife Hunting." 2016.

11. IUCN, SULi, IIED, CEED, Austrian Ministry of Environment, and TRAFFIC, *Beyond Enforcement: Communities, Governance, Incentives and Sustainable Use in Combating Wildlife Crime,* Symposium Report, 26–28 February 2015, Glenburn Lodge, Muldersdrift, South Africa, 2015. Mackenzie, Chapman, and Sengupta, "Spatial Patterns of Illegal Resource Extraction."

12. A. Pyhälä, A. Osuna Orozco, and S. Counsell, *Protected Areas in the Congo Basin: Failing Both People and Biodiversity?* (London: Rainforest Foundation UK, 2016), 80–81.

13. Jerome Lewis, "How Sustainable Development Ravaged the Congo Basin," *Scientific American,* May 2020. https://www.scientificamerican.com/article/how-sustainable-development-ravaged-the-congo-basin/.

14. https://cites.org/prog/iccwc.php/Wildlife-Crime, accessed 2 June 2020.

15. Hauenstein et al., "African Elephant Poaching Rates."

16. Roe, ed., *Conservation, Crime and Comunities;* Roe et al., *Beyond Enforcement.* Lubilo, Rodgers, and Hebink. "Local Hunting."

17. Roe et al., 'Which Components or Attributes of Biodiversity Influence Which Dimensions of Poverty?"; Lubilo, Rodgers and Paul Hebink. "Local Hunting."

18. Roe et al., "Which Components or Attributes of Biodiversity Influence Which Dimensions of Poverty?"

19. Duffy, Emslie, and Knight, "Rhino Poaching,"6.

20. Lunstrum and Givá, "What Drives Commercial Poaching?" Knapp, Peace, and Bechtel, "Poachers and Poverty." Lubilo, Rodgers, and Hebink. "Local Hunting."

21. Pooley, Fa, and Nasi, "No Conservation Silver Lining to Ebola"; Gore et al., "Transnational Environmental Crime."

22. https://www.ecdc.europa.eu/en/geographical-distribution-2019-ncov-cases, accessed 17 June 2021.

23. https://lioncoalition.org/2020/04/04/open-letter-to-world-health-organization/, accessed 10 April 2020.

24. https://theconversation.com/why-shutting-down-chinese-wet-markets-could-be-a-terrible-mistake-130625 (accessed 10 April 2020); for a wider discussion of the different angles, see Briefing by TRAFFIC-International, https://www.traffic.org/publications/reports/wildlife-trade-covid-19-and-zoonotic-disease-risks-shaping-the-response/, accessed 10 April 2020.

25. https://cites.org/prog/iccwc.php/Wildlife-Crime accessed 2 June 2020.

26. Soulé, "What Is Conservation Biology?"

27. Litfin, *Ozone Discourses.*

28. O'Riordan and Jordan, "Precautionary Principle."

29. Duffield, *Development, Security*. Buzan and Wilde, *Security.*

30. Buzan and Wæver. *Regions and Powers*. Buzan and Wilde, *Security.* Floyd, "Can Securitization Theory Be Used in Normative Analysis?" Dalby, *Security and Environmental Change.*

31. For wider discussion of environmental geopolitics approaches, see O'Lear, *Environmental Geopolitics.* For critical geopolitics, see Dalby, *Anthropocene Geopolitics;* Dalby, *Security and Environmental Change.*

32. Massé and Lunstrum, "Accumulation by Securitization."

33. Dalby, *Security and Environmental Change.* Buzan and Wæver, *Regions and Powers.*

34. Wyatt, "Russia's Far East's."

35. Lynch and Stretesky, *Exploring Green Criminology.* Gore, "Global Risks." Wyatt, "Illegal Trade." Nurse and Wyatt, *Wildlife Criminology.*

36. Massé et al., "Conservation and Crime Convergence?"

37. United Nations, "Tackling Illicit Trafficking in Wildlife," accessed 12 September 2017, http://www.un.org/en/ga/search/view_doc.asp?symbol=A/69/L.80.

38. White House, "Executive Order," https://www.govinfo.gov/content/pkg/DCPD-201700106/html/DCPD-201700106.htm.

39. 96 Elephants, home page, accessed 24 March 2017, http://www.96elephants.org/chapters.

40. https://www.wcs.org/96-elephants, accessed 17 June 2020.

41. 96 Elephants, home page, accessed 24 March, 2017, http://www.96elephants.org/chapter-3.

42. See Dickman et al., "Wars over Wildlife."

43. 96 Elephants, home page, accessed 24 March 2017, http://www.96elephants.org/chapter-2.

44. McGuire and Haenlein, "Illusion of Complicity." White, "White Gold of Jihad."

45. 96 Elephants, home page, accessed 24 March 2017, http://www.96elephants.org/chapter-2.

46. Duffield, *Development, Security*.

47. Peluso, "Coercing Conservation." Peluso and Watts, eds., *Violent Environments*. Peluso and Vandergeest, "Political Ecologies of War and Forests." Lunstrum, "Green Militarization." Neumann, *Imposing Wilderness*. Ybarra, "Taming the Jungle."

48. Neumann, "War for Biodiversity."

49. Bocarejo and Ojeda, "Violence."

50. Brockington and Duffy, eds., *Conservation and Capitalism*.

51. McAfee, "Selling Nature to Save It." McAfee, "Nature in the Market-World."

52. Asiyanbi, "Political Ecology of REDD+." Massé and Lunstrum, "Accumulation by Securitization." Bocarejo and Ocada, "Violence."

53. For green security, see Kelly and Ybarra, "Green Security." Ybarra, "Blind Passes." For green militarization, see Lunstrum, "Green Militarization." Marijnen and Verweijen, "Selling Green Militarization." Witter, "Why Militarized Conservation May Be Counter-Productive." For green violence, see Büscher and Ramutsindela, "Green Violence." For green wars, see Büscher and Fletcher, "Under Pressure."

54. Duffy, 'Waging a War." Marijnen and Verweijen, "Selling Green Militarization." Büscher and Ramutsindela, "Green Violence."

55. Verweijen, "Microdynamics Approach to Geographies of Violence."

56. Asiyanbi, "Political Ecology of REDD+."

57. For Kaziranga National Park in India, see Barbora, "Riding the Rhino." For Colombia, see Bocarejo and Ojeda "Conservation and Violence." For Honduras, see Loperena, "Conservation by Racialized Dispossesson." For Laos, see Dwyer, Ingalls, and Baird, "Security Exception." For examples from Guatemala, see Ybarra, "Blind Passes." Ybarra, *Green Wars*.

58. Deutz et al., *Financing Nature*, 13.

59. Perry and Karousakis, *Global Biodiversity Finance*, 3.

60. Felbab-Brown, *Extinction Market*, 11.

61. World Bank, "Global Wildlife Program," accessed 4 July 2018, http://www.worldbank.org/en/topic/environment/brief/global-wildlife-program, and UK government, "Illegal Wildlife Trade," accessed 4 July 2018, https://www.gov.uk/government/news/government-launches-new-plans-to-stamp-out-the-illegal-wildlife-trade-ahead-of-landmark-uk-conference.

62. Interview with 2017_07_17 WP 1.5; World Bank, "Tiger Forum," http://www.worldbank.org/en/news/speech/2010/11/23/remarks-for-first-international-tiger-forum-in-st-petersburg-of-high-level-plenary-session.

63. Wright et al., *Analysis*, 9.

64. Ibid., 14.

65. Global Environment Facility, "Funding to Combat Illegal Wildlife Trade," accessed, 23 November 2018, https://www.thegef.org/news/gef-increases-funding-combat-illegal-wildlife-trade.

66. Ibid.

67. Massé and Margulies, "Geopolitical Ecology of Conservation Assistance."

68. DEFRA, Illegal Wildlife Trade Challenge Project funding, https://assets.publishing.service.gov.uk/government/uploads/system/uploads/attachment_data/file/811381/iwt-project-list-2019.pdf, accessed 5 June 2020.

69. I sat on the Illegal Wildlife Trade Challenge Fund advisory board for three years, 2013–16. I raised this criticism on each round: that the ways the criteria were designed meant that demand reduction and sustainable livelihoods would lose out to law enforcement, and that the requirement to fill out a complex logframe plus the need to submit detailed financial accounts for all partner organizations did not facilitate applications from smaller-scale locally based organizations. Instead, the criteria and application process encouraged, in my view, large-scale conservation NGOs with experience in grant writing for funders from the global North.

70. DEFRA, Illegal Wildlife Trade Challenge Project funding, https://assets.publishing.service.gov.uk/government/uploads/system/uploads/attachment_data/file/811381/iwt-project-list-2019.pdf, accessed 5 June 2020.

71. Interview with 2017_07_12 WP 1.2.

72. Ibid.

73. Holmes, "Biodiversity for Billionaires," 185; also see Ramutsindela, "Extractive Philanthropy."

74. Fortwrangler, "Friends with Money." For a discussion of the influence of celebrities on approaches to and practices of conservation, see Brockington, *Celebrity and the Environment.*

CHAPTER 2. FRAMINGS MATTER

1. CITES, "Wildlife Crime," accessed 14 March 2020, https://cites.org/prog/iccwc.php/Wildlife-Crime.

2. Elliott, "Governing."

3. Gore, "Global Risks." See also Massé et al., "Conservation and Crime Convergence?" Felbab-Brown, *Extinction Market,* 10. Wittig, "Poaching."

4. TRAFFIC "Wildlife Trade," accessed 13 October 2020, http://www.traffic.org/trade/. See also Sina, *Wildlife Crime.*

5. European Commission, "Wildlife Trafficking," accessed 13 October 2020, http://ec.europa.eu/environment/cites/traf_steps_en.htm; TRAFFIC, "Wildlife Trade," accessed 13 October 2020, http://www.traffic

.org/trade/. See also Ayling, "What Sustains Wildlife Crime?" OECD, *Illicit Trade,* 59. European Union, *EU Action Plan.*

6. Nelleman, et al., *Rise of Environmental Crime.*

7. Natusch and Lyons, "Exploited for Pets." Kasterine et al., *Trade in South-East Asian Python Skins.*

8. UNODC, *World Wildlife Crime Report,* 23.

9. Gore, "Global Risks." Also see Lavorgna and Sajeva, "Studying the Online Illegal Trade in Plants." Shelley, *Dark Commerce,* 76. Massé et al., "Conservation and Crime Convergence?"

10. Massé et al., "Conservation and Crime Convergence?," 25.

11. Gore, "Global Risks." 2.

12. Interview with 2017_11_20 WP 1.12a.

13. National Wildlife Crime Unit, "Priorities," accessed 13 March 2019, https://www.nwcu.police.uk/how-do-we-prioritise/priorities/.

14. Interview with 2017_07_17 WP 1.3.

15. Interview with 2017_11_20 WP 1.12a.

16. CITES, "Wildlife Crime," accessed 14 November 2020, https://cites.org/eng/prog/iccwc_new.php.

17. Scanlon, "Reflections on Eight Years as Secretary-General of CITES," accessed 12 April 2018, https://www.cites.org/eng/news/sg/early-reflections-on-eight-years-as-sg-of-cites-2010–2018_0604208.

18. Wildlife Service, "Wildlife Enforcement," accessed 14 November 2020, https://www.fws.gov/southeast/law-enforcement/.

19. McLellan and Allan, *Wildlife Crime Initiative. S*ee also TRAFFIC, *Big Wins.*

20. Wyatt and Cao, *Corruption and Wildlife Trafficking.*

21. Interview with 2017_07_19 WP 1.9.

22. Interview with 2017_11_20 WP 1.12.

23. Interview with 2017_11_27 WP 1.13.

24. EUROPOL, "Glass Eel Traffickers," https://www.europol.europa.eu/newsroom/news/glass-eel-traffickers-earned-more-eur-37-million-illegal-exports-to-asia, accessed 28 April 2020.

25. Sustainable Eel Group, "Stop the Illegal Trade," https://www.sustainableeelgroup.org/stop-the-illegal-trade/, accessed 28 April 2020.

26. Van Uhm, *Illegal Wildlife Trade;* Van Uhm, *Illegal Trade in Wildlife and Harms to the World.*

27. Sina et al., *Wildlife Crime,* 24–27.

28. TRAFFIC, *Big Wins.* OECD, *Illicit Trade,* 69.

29. Sollund and Maher, *Illegal Wildlife Trade.* Wyatt, "Illegal Trade."

30. Chaber et al., "Scale of Illegal Meat Importation."

31. Sina et al., *Wildlife Crime,* 1–27.

32. European Commission, "*Cnemaspis psychedelica,*" accessed 13 October 2019, http://ec.europa.eu/environment/cites/pdf/cop17/Cnemaspis%20psychedelica.pdf. See also IUCN, "CITES Conference," accessed

13 October 2019, https://www.iucn.org/news/iucn-informs-key-decisions-cites-conference-wildlife-trade.

33. Margulies et al., "Illegal Wildlife Trade."

34. CITES, "Species," accessed 8 March 2019, https://www.cites.org/eng/disc/species.php; Lavorgna, et al., "CITES, Wild Plants, and Opportunities for Crime."

35. IUCN Orchid Specialist Group home page, accessed 8 March 2019, http://globalorchidtrade.wixsite.com/home. Lavorgna and Sajeva, "Studying Illegal Online Trades in Plants."

36. Margulies, "Korean Housewives and Hipsters."

37. Van Uhm, *Illegal Wildlife Trade;* Van Uhm, *Illegal Trade in Wildlife and Harms to the World.* Also see Lavorgna and Sajeva, "Studying Illegal Online Trades in Plants."

38. Dickinson, "Black Gold and Grey Areas: Examining the Impacts of Regulations on the Geopolitical-Ecology of Caviar Trade in the European Union" (PhD thesis, University of Sheffield, 2020). Also see UNODC, *World Wildlife Crime Report,* 86–91.

39. Shelley, *Dark Commerce,* 86

40. Lubilo, Rodgers, and Hebinck. "Local Hunting."

41. WWF-International/TRAFFIC-International, *Big Wins in the War against Wildlife Crime.*

42. Lubilo, Rodgers, and Hebinck. "Local Hunting." MacKenzie, *Empire of Nature.* Neumann, "War for Biodiversity." Duffy "Waging a War to Save Biodiversity."

43. See Lubilo, Rodgers, and Hebinck. "Local Hunting." Challender and MacMillan, "Poaching." Duffy, St. John, Büscher, and Brockington, "Poverty and Illegal Wildlife Hunting." Hübschle, "Rhino Poaching." Peluso, "Coercing Conservation."

44. Adams, *Against Extinction.* Bodmer and Lozano, "Rural Development." Fischer et al., "(De)legitimizing Hunting." Mackenzie, *Empire of Nature.*

45. Duffy, St. John, Büscher. and Brockington, "Poverty and Illegal Wildlife Hunting." Sanderson and Redford. "Contested Relationships."

46. Duffy, St. John, Büscher, and Brockington, "Poverty and Illegal Wildlife Hunting."

47. European Union, *EU Action Plan.*

48. "Baobab: Trading the Once Forbidden Fruit," *New Agriculturalist,* January 2009, accessed 1 December 2020, http://www.new-ag.info/en/developments/devItem.php?a=640.

49. "Baobab Products Mozambique," SEED, accessed 17 June 2021, https://www.seed.uno/enterprise-profiles/baobab-products-mozambique.

50. Cooney et al., *Trade in Wildlife.*

51. Ibid., 16. Roe, *Trading Nature.* Clemente-Munoz, "The Role of CITES."

52. Cooney et al., *Trade in Wildlife,* 15. Brashares et al., "Wildlife Decline."

53. Roe, "Despite COVID-19 Using Wild Species May Still Be the Best Way to Save Them," https://www.iied.org/despite-covid-19-using-wild-species-may-still-be-best-way-save-them, accessed 28 April 2020.

54. Gore, "Global Risks." Wyatt, "Corruption."

55. Wyatt, "Illegal Trade."

56. Gore, "Global Risks." For a further discussion, see Lynch and Stretesky, *Exploring Green Criminology.*

57. Sollund, "Wildlife Management, Species Injustice and Ecocide in the Anthropocene." Sollund, *Crimes of Wildlife Trafficking.*

58. Eckersley, "Ecological Intervention," 293. Sollund, "Wildlife Management, Species Injustice and Ecocide in the Anthropocene."

59. Humphreys and Smith. "Rhinofication," 121.

60. Humphreys and Smith, "War and Wildlife," 137–40.

61. Eckersley, "Ecological Intervention," 311–12; also see Cochrane and Cooke, "Humane Intervention."

62. Interview with 2017_11_27 WP 1.13.

63. Gore, "Global Risks."

64. Shelley, *Dark Commerce.*

65. Lavorgna and Sajeva, "Studying Illegal Online Trades in Plants." Felbab-Brown, *Extinction Market,* 38–40.

66. https://www.unodc.org/unodc/en/organized-crime/intro/UNTOC.html#Fulltext, accessed 29 April 2020.

67. UNODC, *World Wildlife Crime Report,* 23. Shelley, *Dark Commerce,* 178.

68. Elliott, "Criminal Networks," 27–28. See also Titeca, "Ivory Trade." Wasser et al., "Conservation."

69. EUROPOL, *Threat Assessment 2013,* accessed 8 March 2019, https://www.europol.europa.eu/activities-services/main-reports/serious-and-organized-crime-threat-assessment.

70. Ibid.

71. UNODC, "Wildlife and Forest Crime," accessed 22 August 2019, https://www.unodc.org/unodc/en/wildlife-and-forest-crime/index.html.

72. Broussard, "United Nations," 465.

73. Ibid., 463.

74. UNODC, *World Wildlife Crime Report*.

75. UNODC, "Wildlife and Forest Crime," accessed 1 December 2020, https://www.unodc.org/unodc/en/wildlife-and-forest-crime/index.html.

76. Van Asch, "International Consortium," 469.

77. CITES, "International Consortium," accessed 13 October 2019, http://www.cites.org/eng/prog/iccwc.php; http://www.Interpol.int/Crime-areas/Environmental-crime/International-Consortium-on-Combating-Wildlife-Crime.

78. Van Asch, "International Consortium," 473.

79. CITES, "Forum," accessed 1 December 2020, https://cites.org/eng/news/pr/CoP17_hosts_first_ever_wildlife_crime_partnerships_forum_26092016.

80. WWF, "Crime Initiative," accessed 1 December 2020, https://wwf.panda.org/discover/our_focus/wildlife_practice/wildlife_trade/wildlife_crime_initiative/.

81. Wyatt, "Corruption." Wyatt and Cao, "Corruption."

82. Elliott, "Criminal Networks," 32–34.

83. Wyatt, "Corruption," 156–58.

84. Sina et al., *Wildlife Crime,* 32–34. Wyatt and Kushner, *Global.* Wyatt, "Illegal Trade." Elliot, "Criminal Networks." Lawson and Vines, *Global Impacts.*

85. INTERPOL, *Elephant Poaching,* 1–2.

86. Titeca, "Ivory Trade." See also Somerville, *Ivory*. UNODC, *Transnational Organized Crime*. Wittig, "Poaching," 95. Haenlein and Keatinge, *Follow the Money,* 18.

87. Massé et al., "Conservation and Crime Convergence?," 31.

88. Wyatt, "Illegal Trade."

89. Felbab-Brown, *Extinction Market,* 40–45.

90. Interview with 2017_07_18 WP 1.6a (c).

91. Interview with 2017_07_18 WP 1.6a (e).

92. Wyatt, "Illegal Trade." Also see Wittig, "Poaching," 95.

93. Titeca, "Ivory Trade," 14. Also see Kassa et al., *Worm's-Eye View of Wildlife Trafficking in Uganda.*

94. Titeca, "Ivory Trade," 15. Also see Rundhovde, "Merely a Transit Country?"

95. See Hübschle, "Security Coordination," 208. Wittig, "Poaching," 95. Haenlein and Keatinge, *Follow the Money,* 18.

96. Massé et al., "Conservation and Crime Convergence?," 31.

97. Felbab-Brown, *Extinction Market,* 19–26.

98. EUROJUST, *Strategic Project*, 14.

99. OECD, *Illicit Trade,* 59.

100. US Congress, National Defense Authorization Act for Fiscal Year 2016, H.R. 1735, https://www.congress.gov/bill/114th-congress/house-bill/1735/text, accessed 18 April 2019.

101. Interview with 2017_07_18 WP 1.6a (e). Also see Shelley, *Dark Commerce,* 139–40.

102. Felbab-Brown, *Extinction Market,* 11.

103. European Commission, *European Agenda on Security*, 4. Nelleman et al., *Rise of Environmental Crime,* 30.

104. Interview with 2017_11_20 WP 1.12.

105. Interview with 2017_11_27 WP 1.13.

106. Lindenmayer and Scheele, "Do Not Publish."

107. To, Mahanty, and Dressler, "Social Networks of Corruption."

108. Anonymous, "Rosewood Democracy." Also see UNODC, *World Wildlife Crime Report,* 33–40.

109. See Williams and Le Billon, eds., *Corruption.*

CHAPTER 3. WAR FOR BIODIVERSITY

1. Büscher and Ramutsindela, "Green Violence."

2. Mogomotsi and Madigelelele, "Live by the Gun, Die by the Gun." Henk, "Botswana Defence Force and the War against Poachers in Southern Africa." Henk, "Biodiversity and the Military in Botswana."

3. Interview with Luis Arranz, WWF, https://www.dw.com/en/conservation-in-times-of-war/a-46719951, accessed 4 December 2020.

4. Hillborn et al., "Effective Enforcement in a Conservation Area."

5. Duffy et al., "Why We Must Question the Militarization of Conservation." Day, "Sister Forces."

6. Bachmann, Bell, and Holmqvist, *War.* Massé. "Conservation Law Enforcement."

7. Wilson and Boratto, "Conservation, Wildlife Crime, and Tough-on-Crime Policies."

8. Marijnen and Verweijen, "Selling Green Militarization," 276–79. Verweijen, "Microdynamics Approach to Geographies of Violence."

9. Virunga the Movie, home page, accessed 28 March 2019, https://virungamovie.com.

10. Fred Kockott. "Twelve Rangers Killed in Latest Virunga Park Incident," *Mongabay,* 25 April 2020, https://news.mongabay.com/2020/04/twelve-rangers-killed-in-latest-virunga-park-incident/, accessed 4 May 2020.

11. Verweijen, "Microdynamics Approach to Geographies of Violence."

12. Thin Greenline, home page, accessed 3 March 2017, http://www.thingreenline.org.au/.

13. United for Wildlife, "The Facts," accessed 12 April 2017, https://www.unitedforwildlife.org/#!/the-facts

14. IUCN, "Rising Murder Toll," accessed 14 March 2017, http://www.iucn.org/?17196/Rising-murder-toll-of-park-rangers-calls-for-tougher-laws.

15. WWF, "Rangers," accessed 14 March 2017, http://www.worldwildlife.org/stories/by-the-numbers-rangers-combat-wildlife-crime.

16. The figure of one thousand rangers was also discussed at the ZSL Symposium on International Wildlife Trafficking in February 2014, which was linked to the London Conference and London Declaration. A recording of the ZSL conference is available at http://www.zsl.org/science/previous-scientific-events/symposium-international-wildlife-trafficking, accessed 14 March 2017. The author was present at the conference.

17. Humphreys and Smith, "Rhinofication." Barbora, "Riding the Rhino." Also see interview with Luis Arranz, https://www.dw.com/en/conservation-in-times-of-war/a-46719951, accessed 4 December 2020.

18. Lunstrum, "Green Militarization."

19. "Masisi Revokes Shoot-to-Kill Policy," *Southern Times,* 1 June 2018, accessed 17 June 2021, https://africasustainableconservation.com/2018/06/01/botswana-masisi-ends-shoot-to-kill-approach-to-conservation/.

20. See Duffy, *Killing for Conservation.* Somerville, *Ivory.*

21. Büscher and Fletcher, "Under Pressure."

22. Adams, *Against Extinction;* Brockington, Duffy, and Igoe, *Nature Unbound.* Jacoby, *Crimes against Nature.*

23. Duffy, *Nature Crime,* 79–112. Neumann, "War for Biodiversity." Leakey, *Wildlife Wars.*

24. Leakey, *Wildlife Wars.* See also Somerville, *Ivory.*

25. Duffy, *Nature Crime,* 113–54. Somerville, *Ivory.*

26. Dressler et al., "Hope to Crisis," 5.

27. Ibid., 7–11. See also Mukamuri et al., eds., *Beyond Proprietorship.*

28. Murombedzi, "Wildlife Conservation."

29. Neumann, "Disciplining Peasants"; *Imposing Wilderness.*

30. Dzingirai, "New Scramble."

31. Mbaria and Ogada, "Big Conservation Lie." See also Bersaglio, "Green Violence."

32. Mbaria and Ogada, "Big Conservation Lie," 160–62. Dressler et al., "Hope to Crisis."

33. See Hutton et al., "Back to the Barriers?" Terborgh, *Requiem for Nature.* Oates, *Myth and Reality.* Brandon, Redford, and Sanderson, *Parks in Peril.*

34. Interview with 2017_07_12 WP 1.2. Also see European Commission, *Study on the Interaction between Security and Wildlife,* 16.

35. Interview with 2020_04_14 WP 1.20a.

36. Interview with 2017_07_17 WP 1.3.

37. Tsavo Trust, "Stabilization through Conservation," accessed 23 March 2015, http://tsavotrust.org/stabilcon/.

38. The links between conservation and stabilization were discussed at the StabilCon launch event at the Royal United Services Institute (RUSI) in London on 1 December 2014. It was attended by Ian Saunders, chief operations officer, Tsavo Trust, Professor Judi Wakhungu, secretary of state for the environment, water, and natural resources, Republic of Kenya; Ambassador Hussein Dado, governor, Tana River County, Republic of Kenya. I was an invited member of the discussion panel.

39. Massé, Gardiner, Lubilo, and Themba, "Inclusive."

40. Interview with 2018_01_23 WP 1.14b.

41. Interview with 2018_01_23 WP 1.14a.

42. Interview with 2017_07_17 WP 1.6b.

43. Interview with 2018_01_23 WP 1.14a.

44. Hartter and Goldman, "Local Responses."

45. Kelly and Gupta, "Protected Areas."

46. Ibid. See also Kelly, "Crumbling Fortress."

47. Kelly and Gupta, "Protected Areas," 175. Kelly, "Crumbling Fortress."

48. Lombard and Tubiana, "Bringing the Tracker-Guards Back In."

49. Büscher and Fletcher, "Under Pressure."

50. Patrick Greenfield and Peter Muiruri. "Conservation in Crisis: Ecotourism Collapse Threatens Communities and Wildlife." *The Guardian*, 5 May 2020, https://www.theguardian.com/environment/2020/may/05/conservation-in-crisis-covid-19-coronavirus-ecotourism-collapse-threatens-communities-and-wildlife-aoe, accessed 10 May 2020.

51. Moore, *Capitalism*. Castree, "Neoliberalism." Bigger and Robertson, "Value." McAfee, "Selling Nature."

52. McAfee, "Contradictory Logic." Büscher and Fletcher, "Accumulation."

53. Sullivan, "Banking Nature." McAfee and Shapiro, "Payments." Holmes and Cavanaugh, "Social Impacts."

54. Kathy McAfee, "Selling Nature." See also McAfee, "Contradictory Logic."

55. Brockington, Duffy, and Igoe, *Nature Unbound*, 195–97. Igoe, *Nature of Spectacle*.

56. Holmes, "Conservation's Friends."

57. Biggs et al., "Legal Trade."

58. Ibid. Duffy, "North-South Dynamics."

59. White House (Obama administration), *Combating Wildlife Trafficking*, Executive Order 13648 of July 1, 2013, Federal Register 78(129), July 5, 2013, pp. 40621–23. See also Wyler and Sheikh, *International Illegal Trade*, 2.

60. Interview with 2017_07_17 WP 1.4.

61. Interview with 2017_07_18 WP 1.6a (e).

62. Full text of the END Wildlife Trafficking Act 2016 is available at https://www.congress.gov/114/plaws/publ231/PLAW-114publ231.pdf, accessed 14 May 2018.

63. Corson, *Corridors*.

64. Interview with 2017_07_17 WP 1.4.

65. Duffy, "War by Conservation."

66. Büscher and Fletcher, "Under Pressure." Marijnen and Verweijen, "Selling Green Militarization." Lombard, "Threat Economies."

67. Interview with 2017_07_17 WP 1.3.

68. Duffy et al., "Militarization." Barichievy et al., "Do Armed Field-Rangers Deter Rhino Poachers?" Duffy et al., "Poverty and Illegal Hunting." Barbora, "Riding the Rhino."

69. For Laos, see Dwyer et al., “Security Exception.” For Guatemala, see Devine, “Counterinsurgecy Ecotourism,” and Ybarra, “Blind Passes.” For Colombia, see Bocarejo and Ojeda, “Violence and Conservation.” For Democratic Republic of the Congo (DRC), see Marijnen and Verweijen, “Selling Green Militarization,” and Verweijen, “Microdynamics Approach.” For Nigeria, see Asiyanbi, “Political Ecology of REDD+.” For India, see Barbora, “Riding the Rhino”; Muralidharan and Rai, “Violent Maritime Spaces”; and Dutta, Forest Becomes Frontline.” For Tanzania, see Mabele, “Beyond.”

70. See Mackenzie, *Empire of Nature.* Hübschle, “Rhino Poaching.” Neumann, “War for Biodiversity.” Peluso “Coercing Conservation.”

71. Duffy et al., “Poverty and Illegal Wildlife Hunting.” Cooney et al., From Poachers to Protectors.” Lunstrum, “Capitalism.”

72. Duffy et al., “Militarization.” Duffy et al., “Poverty and Illegal Wildlife Hunting.” Cooney et al., *Trade in Wildlife.* Hübschle, “Rhino Poaching.” Massé et al., “Inclusive.”

73. Lunstrum, “Green Militarization.” Büscher and Ramutsindela, “Green Violence.”

74. For further discussion see Hübschle, “Rhino Poaching.” Massé and Lunstrum, “Accumulation by securitization.” Massé et al., “Green Militarization.”

75. Interview with 2017_05_26 WP 1.1a.

76. Carlson et al., *In the Line of Fire.* Mabele, “Beyond.” Annecke and Masubelele, “Impact of Militarization.” Kuiper et al., “Ranger Perceptions.” Mushonga. “Militarization of Conservation.”

77. https://www.gameranger.org/what-we-do/ranger-care.html, accessed 30 June 2020.

78. Annecke and Masubelele, “Impact of Militarization.” Massé et al., “Inclusive.” Moreto, “Introducing Intelligence-Led Conservation.” Duffy et al., “Militarization.”

79. WWF, “Ranger Perceptions Asia,” https://c402277.ssl.cf1.rackcdn.com/publications/861/files/original/Ranger_Perception_Survey_%28Asia%29.pdf?1456408489.

80. WWF, “Ranger Perceptions Africa,” https://c402277.ssl.cf1.rackcdn.com/publications/880/files/original/Ranger_Perception_Africa_%28FINAL%29.pdf?1464013250.

81. Lombard and Tubiana, “Bringing the Tracker Guards Back In.”

82. Dutta, “Forest Becomes Frontline.”

83. Interview with 2017_05_26 WP 1.1a.

84. Interview with 2019_11_12 WP 1.19.

85. Lombard and Tubiana, “Bringing the Tracker Guards Back In.”

86. Titeca et al., “Conservation as a Social Contract in a Violent Frontier.” Also see European Commission, *Study on the Interaction between Security and Wildlife,* 25.

87. Interview with 2017_05_26 WP 1.1b.

88. Interview with 2017_05_26 WP 1.1a.

89. Marijnen, De Vries, and Duffy, "Conservation in Violent Environments."

90. Duffy et al., "Militarization."

91. Cooney et al., "From Poachers to Protectors." Annecke and Masubele, "Impact of Militarization." Haas and Ferreira, "Politically Feasible Conservation Policies." Sundström, et al., "Gender Differences in Poaching Attitudes."

92. Lunstrum, "Capitalism."

93. Interview 2017_07_17 WP 1.3.

94. Ibid.

95. Interview 2017_07_18 WP 1.6b (b).

96. Survival International, home page, accessed 3 July 2018, https://www.survivalinternational.org/stopthecon.

97. Survival International, "Conservation," accessed 3 July 2018, https://www.survivalinternational.org/conservation.

98. Ibid.

99. Buzzfeed, "WWF Torture," accessed 7 March 2018, https://www.buzzfeednews.com/article/tomwarren/wwf-world-wide-fund-nature-parks-torture-death.

100. US Government Accountability Office. *Combating Wildlife Trafficking: Agencies Work to Address Human Rights Abuse Allegations in Overseas Conservation Programs* (2020), https://www.gao.gov/products/GAO-21-139R, accessed 2 December 2020.

101. WWF-International. *Embedding Human Rights in Nature Conservation.*

102. https://www.rainforestfoundationuk.org/wwfs-lack-of-contrition-as-independent-review-finds-systemic-failings-in-its-treatment-of-human-rights, accessed 4 December 2020.

103. Pyhälä, Osuna-Orozco, and Counsell, *Protected Areas*. Ayari and Counsell, *Human Cost*.

104. Pyhälä, Osuna-Orozco, and Counsell, *Protected Areas,* 80–81. For further discussion, see Twinamatsiko et al., *Linking Conservation*.

105. Pyhälä, Osuna-Orozco, and Counsell, *Protected Areas,* 80–81. IUCN et al., *Beyond Enforcement*.

106. Interview with 2018_01_23 WP 1.14a.

107. Ibid. Also see Day, "Sister Forces."

CHAPTER 4. TERRORISM AND POACHING

1. Peluso, "Coercing Conservation," 205–9.

2. Duffy, "War by Conservation."

3. Sollund and Maher, *Illegal Wildlife Trade,* 1.

4. IFAW, "Criminal Nature," accessed 17 June 2021, https://www.ifaw.org/resources/criminal-nature-the-global-security-implications-of-the-illegal-wildlife-trade.

5. Vira and Ewing, *Ivory's Curse.*

6. Interview with 2019_11_12 WP 1.19.

7. Interview with 2017_11_27 WP 1.13.

8. European Commission, *Study on the Interaction between Security and Wildlife,* 20.

9. Wittig, *Understanding.*

10. Duffy, "Waging a War to Save Biodiversity." White, "White Gold of Jihad." Lawson and Vines, *Global Impacts.* Duffy, "War by Conservation."

11. Sollund and Maher, *Illegal Wildlife Trade,* 2.

12. European Commission, *Study on the Interaction between Security and Wildlife, 27.*

13. Interview with 2017_11_ 13 WP 1.13c.

14. Interview with 2017_07_18 WP 1.6a (c).

15. Wittig, *Understanding.*

16. See Ratchford, Allgood, and Todd, *Criminal Nature,* 12–13. Somerville, *Ivory.* Lombard, Threat Economies.

17. Conservation International, "Global Stability," accessed 14 August 2014, http://www.conservation.org/what/pages/global-stability.aspx

18. Conservation International, "Promoting Economic, National and Global Security," accessed 14 August 2014, http://www.conservation.org/projects/Pages/Promoting-Economic-National-and-Global-Security-Direct-connection.aspx.

19. Seligmann, "One Way to Fight Terrorism: End the Ivory Trade," accessed 25 March 2014, http://blog.conservation.org/2013/10/one-way-to-fight-terrorism-end-the-ivory-trade/.

20. Brockington, *Celebrity and the Environment.* Büscher et al., "Towards a Synthesized Critique."

21. Keating, "Ivory Funds Terrorism?" accessed 13 August 2014, http://www.slate.com/blogs/the_world_/2013/10/02/is_the_illegal_ivory_trade_funding_terrorist_groups_like_al_shabab.html.

22. Wildlife Conservation Society, "96 Elephants Campaign," accessed 25 March 2014, http://96elephants.org/. See also Wildlife Conservation Society, "Press release," accessed 14 August 2014, http://www.wcs.org/press/press-releases/96-elephants.aspx.

23. Wildlife Conservation Society, "96 Elephants Campaign," accessed 13 August 2014, http://www.96elephants.org/chapter-2.

24. EAL, 2012.

25. Duffy, "War by Conservation."

26. *The Independent,* "Elephant Campaign: How Africa's 'White Gold' Funds the al-Shabaab Militants," accessed 25 June 2014, http://www.independent.co.uk/voices/campaigns/elephant-campaign/elephant-

campaign-how-africas-white-gold-funds-the-alshabaab-militants-9102862.html.

27. Somerville, *Ivory,* 309. Lawson and Vines, *Global Impacts.*

28. Ratchford, Allgood and Todd, *Criminal Nature.*

29. Ibid., 12–13.

30. Laurel Neme, Andrea Crosta and Nir Kalron, "Terrorism and the Ivory Trade," *Los Angeles Times,* 14 October 2013.

31. Neme, Crosta and Kalron, "Al-Shabaab and the Human Toll of the Illegal Ivory Trade," *National Geographic,* 3 October 2013

32. *The Independent,* "Elephant Campaign: How Africa's 'White Gold' Funds the al-Shabaab Militants," accessed 25 June 2014, http://www.independent.co.uk/voices/campaigns/elephant-campaign/elephant-campaign-how-africas-white-gold-funds-the-alshabaab-militants-9102862.html.

33. Somerville, *Ivory,* 308–11.

34. See http://www.lastdaysofivory.com/, accessed 23 May 2015.

35. Personal communication from conservation professional via email to author (25 February 2014), http://elephantleague.org/project/africas-white-gold-of-jihad-al-shabaab-and-conflict-ivory/.

36. McGuire and Haenlein, *An Illusion of Complicity,* 19.

37. Somerville, *Ivory,* 308–10.

38. McGuire and Haenlein, *An Illusion of Complicity.* See also Duffy, "War by Conservation." McGuire, "Kenya's War on Poaching."

39. Leakey, *Wildlife Wars,* 102.

40. Duffy, "War by Conservation."

41. Interview with 2017_07_17 WP 1.5c.

42. Interview with 2017_07_17 WP 1.5c.

43. Somerville, *Ivory,* 311.

44. Duffy, "War by Conservation."

45. Interview with 2017_07_17 WP 1.3.

46. Interviews with 2017_11_ 13 WP 1.13c; 2017_11_ 13 WP 1.13b.

47. Interview with 2019_11_12 WP 1.19.

48. Transcript of expert witness evidence provided at a hearing of ICCF, http://iccfoundation.us/index.php?option=com_content&view=article&id=447&Itemid=369

49. ICC, home page, accessed 29 April 2018, http://iccfoundation.us/index.php?option=com_content&view=article&id=447&Itemid=369.

50. ICC, home page, accessed 29 April 2018, http://iccfoundation.us/index.php?option=com_content&view=article&id=447&Itemid=369.

51. ICC, home page, accessed 13 August 2018, http://iccfoundation.us/index.php?option=com_content&view=article&id=445&Itemid=367.

52. Christy, "Blood Ivory, Ivory Worship," *National Geographic,* accessed 15 August 2018, http://ngm.nationalgeographic.com/2012/10/ivory/christy-text

53. Gettleman, "Elephants Dying in an Epic Frenzy as Ivory Fuels Wars and Profits," *New York Times,* http://www.nytimes.com/2012/09/04/world/africa/africas-elephants-are-being-slaughtered-in-poaching-frenzy.html?pagewanted=all&_r=0.

54. White, "White Gold of Jihad."

55. ICC, "Hearing," accessed 29 April 2014, http://iccfoundation.us/downloads/Hearing_Testimony_Saunders.pdf.

56. John Scanlon, expert witness testimonial to the US Senate Foreign Relations Committee Hearing "Ivory and Insecurity: The Global Implications of Poaching in Africa," 22 May 2012, accessed 15 August 2018, http://www.foreign.senate.gov/imo/media/doc/Scanlon_Testimony.pdf.

57. *The Guardian,* "Wildlife Crime," accessed 6 February 2019, https://www.theguardian.com/environment/2013/mar/01/people-animals-wildlife-crime.

58. ICC, "Hearing," accessed 29 April 2018, 4, http://iccfoundation.us/downloads/Hearing_Testimony_Fay.pdf.

59. Interview with 2017_07_17 WP 1.4.

60. White, "White Gold of Jihad." See also Ratchford, Allgood, and Todd, *Criminal Nature,* 25.

61. Medina, "White Gold of Jihad," *New York Times,* 30 September 2013, accessed 13 August 2018, http://www.nytimes.com/2013/10/01/opinion/the-white-gold-of-jihad.html?_r=0.

62. McGuire and Haenlein, *Illusion of Complicity,* 6.

63. AWF, "Partnership," http://www.awf.org/news/conservation-partners-announce-80m-clinton-global-initiative-commitment-action-partnership-save.

64. WCS, "Clinton Global Initiative Commitment to Action," accessed 30 April 2018, http://www.wcs.org/press/press-releases/african-elephants-get-major-boost.aspx. See also Sellar, *Lone Ranger,* 73–76.

65. Nellemann et al., eds., *Environmental Crime Crisis,* 78–81. McGuire and Haenlein, *Illusion of Complicity.*

66. Menkhaus, cited in Williams, "After Westgate," 909.

67. McGuire and Haenlein, *Illusion of Complicity.*

68. Interview with 2017_11_ 13 WP 1.13c.

69. UNEP et al., *African Elephant Crisis,* 12.

70. European Commission, *Study on the Interaction between Security and Wildlife, 4.*

71. Ratchford, Allgood, and Todd, *Criminal Nature,* 12–13.

72. Nellemann et al., eds., *Environmental Crime Crisis,* 78–81. Sellar, *Lone Ranger,* 73–76.

73. European Commission, *Study on the Interaction between Security and Wildlife, 25.*

74. Interview with 2017_05_26 WP 1.1a; interview with 2017_05_26 WP 1.1b; interview with 2017_07_12 WP 1.2. Also see Titeca and Edmond, "Out of the Frame."

75. Partners are Zoological Society of London, Wildlife Conservation Society, Conservation International, Flora and Fauna International, WWF-International, IUCN, and the Nature Conservancy. See http://www.unitedforwildlife.org, accessed 14 August 2020.

76. United for Wildlife, "Conflict," accessed 14 August 2020, http://www.unitedforwildlife.org/#!/the-facts/conflict.

77. "Prince Charles Calls for a War on Animal Poachers," *The Guardian,* 21 May 2013, http://www.theguardian.com/environment/2013/may/21/prince-charles-war-animal-poachers.

78. https://www.gov.uk/government/collections/illegal-wildlife-trade-iwt-challenge-fund, accessed 12 January 2021.

79. Titeca, "Out of Garamba." Titeca and Costeur, "L.R.A."

80. Somerville, *Ivory,* 263; and interview with 2017_05_26 WP 1.1a; interview with 2017_05_26 WP 1.1b; interview with 2017_07_12 WP 1.2.

81. Interview with 2017_05_26 WP 1.1a.

82. Interview with 2017_05_26 WP 1.1b.

83. Interview with 2019_11_12 WP 1.19.

84. Interview with 2017_07_18 WP 1.6b (b).

85. Somerville, *Ivory,* 261; European Commission, *Study on the Interaction between Security and Wildlife,* 25.

86. Titeca and Costeur, "LRA." Titeca and Edmond, "Out of the Frame."

87. Interview with 2017_07_18 WP 1.6a (c).

89. Vira and Ewing, *Ivory's Curse,* 11.

89. Kelly-Pennaz, Ahmadou, Moritz, and Scholte, "Cattle."

90. Interview with 2017_07_18 WP 1.6b (b).

91. Interview with 2017_05_26 WP 1.1a.

92. Interview with 2017_07_17 WP 1.3.

93. Interview with 2017_07_17 WP 1.5b.

94. Duffy, "War by Conservation."

CHAPTER 5. SURVEILLANCE, INTELLIGENCE, AND CONSERVATION

1. Eventbrite, "Global Security Event," accessed 6 February 2019, https://www.eventbrite.com/e/criminal-nature-the-global-security-implications-of-illegal-wildlife-trade-tickets-21207843230?aff=ebapi.

2. Written comment from 2020_05_21 WP 1.21.

3. Interview with 2017_07_12 WP 1.2.

4. Maguire, "Kenya's War on Poaching."

5. KWS, "Wildlife Security," accessed 20 August 2018, http://www.kws.go.ke/content/wildlife-security.

6. Humphreys and Smith, "Militarized Responses." Maguire, "Kenya's War on Poaching."

7. Interview with 2020_04_13 WP 1.20b.

8. Kilcullen, "Counter Insurgecy *Redux*." Verweijen and Marijnen. "Counterinsurgency/Conservation Nexus."

9. Kilcullen, "Counter Insurgecy *Redux*."

10. Maguire, "Kenya's War on Poaching."

11. Interview with 2020_04_13 WP 1.20b.

12. Interview (in writing) 2020_06_10 WP 1.22.

13. Maguire, "Kenya's War on Poaching."

14. Humphreys and Smith, "Militarized Responses."

15. 51 Degrees, home page, accessed 4 September 2018, https://www.51degreesltd.com/home.

16. Maguire, "Kenya's War on Poaching." Mbaria and Ogada, "*Big Conservation Lie*," 72.

17. Interview with 2017_07_12 WP 1.2.

18. Mbaria and Ogada, "*Big Conservation Lie*," 72–73.

19. Mkutu and Wandera, "Policing the Periphery."

20. Maguire, "Kenya's War on Poaching."

21. Humphreys and Smith, "Militarized Responses." Maguire, "Kenya's War on Poaching."

22. Maguire, "Kenya's War on Poaching."

23. Mbaria and Ogada, "*Big Conservation Lie*," 69–71. See also Bersaglio, "Green Violence." Bersaglio and Cleaver, "Green Grab."

24. Interview with 2017_07_12 WP1.2.

25. Rademeyer, *Killing for Profit*, 63–99. Hanks, *Operation Lock*.

26. Hanks, *Operation Lock*. See also Spierenburg and Wels, "Conservative Philanthropists."

27. Hanks, *Operation Lock*. Rademeyer, *Killing for Profit*, 63–99.

28. Rademeyer, *Killing for Profit*, 77.

29. Ellis, "Of Elephants and Men."

30. Rademeyer, *Killing for Profit*, 90–99. Ellis, "Of Elephants and Men."

31. Hanks, *Operation Lock*.

32. Ibid., 173.

33. Massé, Lunstrum, and Holterman, "Green Militarization."

34. UNEP, *UNEP Yearbook*.

35. Sellar, *Lone Ranger*, 65.

36. Ibid., 66.

37. Ibid., 42–43.

38. UNEP, "Trade," accessed 13 October 2016, http://www.unep-wcmc.org/citestrade.

39. EU-TWIX, "Database," accessed 4 August 2016, http://www.eutwix.org/.

40. Sacre, "EU-TWIX," 480–81.

41. EUROPOL, "Policy Cycle," accessed 17 September 2018, https://www.europol.europa.eu/crime-areas-and-trends/eu-policy-cycle-empact; EUROPOL, "Illicit Traffic," accessed 17 September 2018, https://www.europol.europa.eu/crime-areas-and-trends/crime-areas/environmental-crime/illicit-trafficking-in-endangered-animal-species.

42. EnviCrimeNet, home page, accessed 17 September 2018, http://www.envicrimenet.eu

43. EUROPOL, "Environmental Crime," accessed 17 September 2018, https://www.europol.europa.eu/crime-areas-and-trends/crime-areas/environmental-crime.

44. Interview with 2017_11_27 WP 1.13.

45. Ibid.

46. Ibid.

47. Written comments from 2020_06_10 WP 1.22.

48. Tittensor, "Evaluating the Relationships between the Legal and Illegal International Wildlife Trades."

49. TRAFFIC-International, *Recent Significant Wildlife Crime-Related Seizures and Prosecutions*.

50. https://eia-international.org/wildlife/wildlife-trade-maps/large-scale-elephant-ivory-seizures/, accessed 18 May 2020.

51. Tittensor, "Evaluating the Relationships between the Legal and Illegal International Wildlife Trades," 5.

52. https://eia-international.org/wildlife/wildlife-trade-maps/illegal-trade-seizures-pangolins/, accessed 18 May 2020.

53. Sellar, *Lone Ranger,* 65.

54. Haenlein and Keatinge, *Follow the Money,* 15.

55. Ibid., 26.

56. IFAW, "Working Together," accessed 13 October 2018, http://www.ifaw.org/united-kingdom/our-work/wildlife-trade/ifaw-and-interpol-working-together-fight-wildlife-crime.

57. EUROPOL, "Join Forces," accessed 17 September 2018, https://www.europol.europa.eu/newsroom/news/europol-and-traffic-join-forces-to-fight-environmental-crime.

58. EUROPOL, "Team against Environmental Crime," accessed 17 September 2018, https://www.europol.europa.eu/newsroom/news/europol-and-wildlife-justice-commission-team-against-environmental-crime.

59. Wildlife Justice, "Analysis," accessed 27 September 2018, https://wildlifejustice.org/intelligence-analysis/.

60. Wildlife Justice, "About Us," accessed 27 September 2018, https://wildlifejustice.org/who-we-work-with/.

61. EUROPOL, "Team against Environmental Crime," accessed 17 September 2018, https://www.europol.europa.eu/newsroom/news/europol-and-wildlife-justice-commission-team-against-environmental-crime, accessed 17 September 2018.

62. Interview with 2017_11_27 WP 1.13.

63. Ibid.

64. Written comments from 2020_06_10 WP 1.22.

65. Interview with 2017_11_20 WP 1.12.

66. Eagle Enforcement, home page, accessed 18 September 2018, http://www.eagle-enforcement.org.

67. Eagle Enforcement, "Operations," accessed 18 September 2018, http://www.eagle-enforcement.org/arrest-operations/.

68. IFAW, "Anti-Poaching," accessed 29 August 2018, https://secure.ifaw.org/united-states/secure/tenboma-anti-poaching-revolution.

69. IFAW, "Conservation," accessed 29 August 2018, https://www.ifaw.org/international/about-us/conservation/faye-cuevas-esq.

70. IFAW, "Conservation," accessed 29 August 2018, https://www.ifaw.org/international/about-us/conservation/faye-cuevas-esq.

71. Interview with 2020_04_13 WP 1.20a.

72. Pink, "INTERPOL," 444–49.

73. EUROPOL, 2013. Sina et al., *Wildlife Crime,* 47.

74. EUROJUST, *Strategic Project,* 12. EUROPOL, *Threat Assessment 2013.*

75. Interview with 2017_11_20 WP 1.12; and interview with 2017_11_27 WP 1.13. EUROPOL, "Policy Cycle," accessed 17 September 2018, https://www.europol.europa.eu/crime-areas-and-trends/eu-policy-cycle-empact.

76. Interview with 2017_11_27 WP 1.13.

77. Interview with 2017_11_20 WP 1.12.

78. Interview with 2017_11_27 WP 1.13.

79. https://biosecproject.org/2020/02/11/news-event-report-iwt-in-europe-wild-birds-caviar/, accessed 19 May 2020.

80. Levi, "Green with Envy," 191.

81. Haenlein and Keatinge, *Follow the Money,* 6.

82. Massé et al., "Conservation and Crime Convergence?," 33

83. Sellar, *Lone Ranger,* 70.

84. See also Haelein and Keatinge, *Follow the Money.*

85. See also ICCWC, Forest and Wildlife Crime Toolkit (rev. ed.), https://www.unodc.org/documents/Wildlife/Toolkit_e.pdf, accessed 29 March 2019.

86. Wittig, "Poaching," 83.

87. Levi, "Green with Envy."

88. Van Uhm, *Illegal Wildlife Trade;* also see Tittensor, "Evaluating the Relationships between the Legal and Illegal International Wildlife Trades."

89. Van Uhm, "Wildlife and Laundering."

90. FATF, "About Us," accessed 25 September 2018, http://www.fatf-gafi.org/about/whoweare/#d.en.11232.

91. Levi, "Green with Envy," 179.

92. Ibid., 183; also see FATF, "Money Laundering and the Illegal Wildlife Trade."

93. Haenlein and Keatinge, *Follow the Money.*

94. Ibid., 8.

95. United Nations, *Report.* UNEP, *UNEP Yearbook,* 25–29.

96. United Nations, 2015 Draft Doha Declaration on Integrating Crime Prevention and Criminal Justice into the Wider United Nations Agenda to Address Social and Economic Challenges and to Promote the Rule of Law at the National and International Levels, and Public Participation, Paragraph 9(e).

97. Wittig, "Poaching," 80.

CHAPTER 6. SECURITY TECHNOLOGIES AND BIODIVERSITY CONSERVATION

1. https://seashepherd.org/news/o-r-c-a-force-uav-captures-first-footage-of-sealers/, accessed 10 March 2020.

2. Gabrys, 2016 Program Earth.

3. Interview with 17_07_27 WP 1.3; also see Laure Joanny, "Law Enforcement Technologies and the Government of Conservation from Indonesian Protected Areas to International Conferences" (PhD thesis, University of Sheffield, 2021).

4. A recent example is the *Ecomodernist Manifesto,* accessed 15 March 2019, http://www.ecomodernism.org.

5. Dunlap and Jakobsen, "*Violent Technologies of Extraction.*"

6. American Geosciences, "Rare Elements," accessed 15 March 2019, https://www.americangeosciences.org/critical-issues/faq/what-are-rare-earth-elements-and-why-are-they-important.

7. Royal Society, *Science.*

8. Black Dot Solutions, home page, accessed 9 November 2018, https://blackdotsolutions.com.

9. Royal Society, *Science,* 23.

10. Di Minin et al., "Use of Machine Learning."

11. Wearn, Freeman, and Jacoby. "Responsible AI for Conservation."

12. Amoore and Raley, "Algorithms," 7. Also see Eubanks, *Automating Inequality.*

13. Amoore and Raley, "Algorithms," 6. Also see Eubanks, *Automating Inequality.*

14. Royal Society, *Science,* 23.

15. Lally, "Crowdsourced," 73.

16. For an overview, see Amoore and Raley, "Algorithms."

17. WCO, "Customs-Environment," accessed 17 June 2021, http://ww\w.wcoomd.org/en/about-us/what-is-the-wco/customs-environment.aspx.

18. Haysom, *Digitally Enhanced Responses,* 2

19. Sollund, and Maher, *Illegal Wildlife Trade*. UNEP, *UNEP Yearbook*, 26–29. Venturini and Roberts, "Disguising Elephant Ivory."

20. Haysom, *Digitally Enhanced Responses,* 3–10.

21. Alfino and Roberts, "Code Word Usage." Also see Venturini and Roberts, "Disguising Elephant Ivory."

22. Hinsley et al., "Estimating." Haysom, *Digitally Enhanced Responses,* 10.

23. Sina et al., *Wildlife Crime,* 37. Hernandez-Castro and Roberts, "Automatic." Milliken, Shaw, Emslie, Taylor, and Turton, *South Africa–Vietnam Rhino Horn Trade Nexus*. Hinsley et al., "Estimating."

24. Hastie and McCrea-Steel, *Wanted,* 5.

25. Ibid., 14.

26. Ibid., 14.

27. Jiao et al., "Greater China," 256.

28. Yu and Jia, *Moving Targets,* 1–3.

29. WeChat, home page, accessed 2 August 2016, http://www.wechat.com/en/.

30. Alibaba, home page, accessed 2 August 2016, http://www.alibaba.com.

31. IFAW, "Reducing Markets," accessed 20 October 2016, http://www.ifaw.org/international/our-work/wildlife-trade/reducing-markets-wildlife-products-china.

32. Yu and Jia, *Moving Targets,* 6.

33. Haysom, *Digitally Enhanced Responses,* 11.

34. IFAW et al., *Global Wildlife,* 10.

35. Sandbrook, Luque-Lora, and Adams, "Human by Catch," 494. Adams, "Geographies of Conservation II." Lunstrum, "Green Militarization." Massé, "Topographies."

36. Interview with 2017_07_20 WP 1.10d.

37. Ibid.

38. Interview with 2017_07_27 WP 1.3.

39. Rogers, "Security." See also Gustafson, Sandstrom, and Townsend, "Bush War," 282.

40. Massé, "Topographies," 58–59.

41. Ibid., 59–60. See also Lunstrum, "Green Militarization."

42. Mulero-Pázmány et al., "Unmanned Aircraft."

43. Ibid.

44. Harrop, "Wild Animal Welfare."

45. Zoohackathon, home page, accessed 22 November 2018, http://www.zoohackathon.com.

46. BIAZA, "Prize," accessed 22 November 2018, https://biaza.org.uk/news/detail/team-odinn-takes-the-price-at-zoohackathon; Zoohackathon, "Past Events," accessed 22 November 2018, http://www.zoohackathon.com/pastevents/2017.

47. Royal Society, *Science,* 11.

48. Sandbrook, Luque-Lora, and Adams, "Human by Catch." See also Massé, "Topographies," 57.

49. Sandbrook, Luque-Lora, and Adams, "Human by Catch," 497. See also Sandbrook, "Drones."

50. Shrestha and Lapeyre, "Modern Wildlife Monitoring."

51. New Scientist, "Imprisoned Conservationist," accessed 8 February 2019, https://www.newscientist.com/article/2184913-irans-imprisoned-conservationists-need-scientists-to-speak-up/.

52. *The Guardian,* "Wildlife Activists," accessed 8 February 2019, https://www.theguardian.com/world/2018/oct/24/iran-charges-five-wild-life-activists-capital-offences-spying.

53. Sandbrook, Luque-Lora, and Adams, "Human by Catch," 500–501.

54. Koh and Wich, "Dawn of Drone."

55. Marris, "Drones in Science."

56. Hollings et al., "Green Sheep?"

57. WWF, "Technology Project," accessed 19 October 2018, https://www.worldwildlife.org/projects/wildlife-crime-technology-project.

58. Ibid.

59. KCAA, "Drone Regulation," accessed 19 October 2018, https://www.kcaa.or.ke/index.php?option=com_content&view=article&id=207:drones-regulations&catid=92:newsandevents&Itemid=742.

60. WWF, "Technology Project," accessed 19 October 2018, https://www.worldwildlife.org/projects/wildlife-crime-technology-project.

61. Millner, "As the Drone Flies," 1. Also see Koh and Wich, "Dawn of Drone Ecology."

62. Gregory, "View to a Kill."

63. Hollings et al., "Green Sheep?," 882–83.

64. Interview with 2018_12_20 WP 1.17.

65. Sandbrook, "Drones."

66. Sandbrook, Luque-Lora, and Adams, "Human by Catch."

67. Massé, "Topographies," 61–62.

68. Gustafson, Sandstrom, and Townsend, "Bush War," 282.

69. Sierra Club, "Poachers vs. Drones," accessed 22 November 2018, https://www.sierraclub.org/sierra/2014–3-may-june/green-life/poachers-v-drones-next-frontier.

70. Snitch, "Satellites, Mathematics and Drones Take Down Poachers in Africa," *The Conversation,* retrieved from https://theconversation.com/satellites-mathematics-and-drones-take-down-poachers-in-africa-36638.

71. Gregory "View to a Kill."

72. Wall, "Ordinary Emergency." Sandbrook, Luque-Lora, and Adams, "Human by Catch."

73. Demmers and Gould. "Assemblage Approach to Liquid Warfare."

74. Massé, "Topographies," 62.

75. Millner, "As the Dronc Flies," 1. Ybarra, "Green Wars." Ybarra, "Taming the Jungle."

76. Massé and Lunstrum, "Accumulation by Securitization."

77. Wildlife Crime Tech, "Winners," accessed 17 January 2019, https://www.wildlifecrimetech.org/winners.

78. Interview with 2017_07_19 WP 1.8.

79. Holmes, "Biodiversity for Billionaires."

80. Wright et al., *Analysis,* 9.

81. Vulcan, "Initiatives," accessed 23 November 2018, http://www.vulcan.com/areas-of-practice/philanthropy/key-initiatives/conservation.

82. This was a central concern raised by several conservation professionals in a closed and confidential knowledge exchange workshop run by the BIOSEC research team. Concerns were expressed that anti-poaching teams on the ground felt pressured to produce figures on interception, arrest, detention, or even killings.

83. Holmes, "Biodiversity for Billionaires." Holmes, "Conservation's Friends in High Places."

84. Millner, "As the Drone Flies."

85. Interview with 2018_12_20 WP 1.17.

86. Kleinschmit et al., *Illegal Logging,* 91.

87. Wainwright, *Decolonizing Development,* 241–57.

88. Millner, "As the Drone Flies."

88. Boekhout van Solinge, "Deforestation."

90. Rainforest Foundation, home page, accessed 23 November 2018, https://www.rainforestfoundationuk.org/rtm.

91. Rainforest Foundation, "Mapping," accessed 23 November 2018, https://www.rainforestfoundationuk.org/mappingforrights.

92. See also Lewis, "Making the Invisible Visible."

93. Lewis. "Making the Invisible Visible." Also see Lewis, "How 'Sustainable Development' Ravaged the Congo Basin."

94. https://www.survivalinternational.org, accessed 20 May 2020. https://www.rainforestfoundationuk.org, accessed 20 May 2020. https://www.iccaconsortium.org/, accessed 20 May 2020.

CHAPTER 7. MILITARY-SECURITY-CONSERVATION NEXUS

1. Mbaria and Ogada, "Big Conservation Lie," 39.

2. Rademeyer, *Killing for Profit,* 87.

3. Devine, "Counterinsurgency Ecotourism."

4. Li and Male, "Conservation of Defense."

5. Lunstrum, "Conservation."

6. Ibid.

7. BIOT, "Marine Protected Area," accessed 21 December 2018, https://biot.gov.io/environment/marine-protected-area/.

8. Jeffrey, *Chagos*.

9. Interview with 2018_11_21 WP 1.16. TUSK, "Poaching," accessed 13 December 2018, https://www.tusk.org/news/22-nov-2017-british-army-helps-reduce-poaching-in-malawi/; *The Guardian*, "Prince Charles Calls for a War on Animal Poachers," accessed 21 May 2013, http://www.theguardian.com/environment/2013/may/21/prince-charles-war-animal-poachers.

10. PWCF, "Illegal Wildlife Trade," accessed 13 December 2018, https://www.pwcf.org.uk/beneficiaries/major-grants-beneficiaries/pwcf-tackles-illegal-wildlife-trade.

11. British Army, "Regiments," accessed 23 November 2017, http://www.army.mod.uk/infantry/regiments/23987.aspx; interview with 2017_11_13 WP 1.11b.

12. African Parks, "Liwonde," https://www.african-parks.org/the-parks/liwonde.

13. Interview with 2018_11_21 WP 1.16.

14. UK Government, "Army Combats Wildlife Trade," accessed 22 November 2017, https://www.gov.uk/government/news/the-british-army-combat-the-illegal-trade-in-wildlife-by-partnering-with-african-parks-and-the-malawian-department-of-national-parks-and-wildlife.

15. MoD and FCO, *UK's International Defence*.

16. UK Government, "British Forces Support Gabon," accessed 30 October 2017, https://www.gov.uk/government/news/british-forces-support-gabons-fight-against-elephant-poachers.

17. British Army, "Regiments," accessed 1 November 2017, http://www.army.mod.uk/infantry/regiments/35343.aspx; UK Government, "British Forces Support Gabon," accessed 30 October 2017, https://www.gov.uk/government/news/british-forces-support-gabons-fight-against-elephant-poachers.

18. Interview with 2018_11_21 WP 1.16.

19. Ibid.

20. Interview with 2017_11_13 WP 1.11b.

21. UK Government, "Army Combats Wildlife Trade," accessed 22 November 2017, https://www.gov.uk/government/news/the-british-army-combat-the-illegal-trade-in-wildlife-by-partnering-with-african-parks-and-the-malawian-department-of-national-parks-and-wildlife.

22. Interview with 2018_11_21 WP 1.16.

23. *The Telegraph*, "New Army Specialists Hunt African Wildlife Poachers and Revive Tracking Skills," accessed 23 November 2018, http://www.telegraph.co.uk/news/2017/08/14/new-army-specialists-hunt-african-wildlife-poachers-revive-tracking/.

24. UK Government, "British Forces Support Gabon," accessed 30 October 2017, https://www.gov.uk/government/news/british-forces-support-gabons-fight-against-elephant-poachers.

25. *The Independent,* "British Troops Tackling Elephant Poachers Selling Ivory to Fund Terror," accessed 20 November 2018, http://www.independent.co.uk/news/world/africa/british-troops-elephant-poachers-ivory-terror-africa-gabon-boko-haram-jihadists-extremists-a7820236.html.

26. *The Mirror,* "Elite British Troops on the Trail of Elephant Killers Who Slaughter Entire Herds for Ivory to Fund Terror," accessed 20 November 2018, http://www.mirror.co.uk/news/world-news/elite-british-troops-trail-elephant-10721620.

27. See Titeca and Edmond, "Outside the Frame" for a discussion of this issue with regards to alleged LRA involvement in ivory trafficking.

28. UK Government, "British Forces Support Gabon," accessed 30 October 2017, https://www.gov.uk/government/news/british-forces-support-gabons-fight-against-elephant-poachers. See also British Army, "Regiments," accessed 1 November 2018, http://www.army.mod.uk/infantry/regiments/43049.aspx.

29. *The Independent,* "Malawi Turns to British Troops in Poaching War," accessed 2 October 2018, https://www.independent.co.ug/malawi-turns-british-troops-poaching-war/.

30. Interview with 2018_11_21 WP 1.16.

31. Interview with 2017_11_13 WP 1.11b.

32. Ibid.

33. BBC News, "British Army on Elephant Saving Mission," accessed 21 November 2017, https://www.youtube.com/watch?time_continue=13&v=bh3G9ViZG24.

34. British Army, "British Army Training in Gabon," https://modmedia.blog.gov.uk/2015/09/28/british-army-anti-elephant-poaching-training-up-and-running-in-gabon/.

35. UK Government, "Army Combats Wildlife Trade," accessed 23 November 2017, https://www.gov.uk/government/news/the-british-army-combat-the-illegal-trade-in-wildlife-by-partnering-with-african-parks-and-the-malawian-department-of-national-parks-and-wildlife.

36. *Daily Mail,* "Malawi Turns to British Troops in Poaching War," accessed 2 November 2017, http://www.dailymail.co.uk/wires/afp/article-5041551/Malawi-turns-British-troops-poaching-war.html; *The Independent,* "Malawi Turns to British Troops in Poaching War," accessed 2 November 2017, https://www.independent.co.ug/malawi-turns-british-troops-poaching-war/.

37. Neumann, "War for Biodiversity."

38. Marijnen and Verweijen, "Selling Green Militarization."

39. Lombard, "Threat Economies." Also see Lombard and Tubiana "Bringing the Tracker-Guards Back In."

40. Neumann, "War for Biodiversity."

41. Rhinos and elephants were reintroduced to Majete National Park in 2008, but since then there have been no incidences of poaching, see https://www.africanparks.org/the-parks/majete, accessed 30 June 2020.

42. Pattison, "War Theory"; *Morality.*

43. See Humphreys and Smith, "Rhinofication," 801.

44. For a wider discussion of PMCs in international interventions, see Pattison, "War Theory"; *Morality.*

45. IAPF, home page, accessed 27 March 2018, http://www.iapf.org.

46. *Akashinga: The Brave Ones,* https://films.nationalgeographic.com/akashinga accessed 28 May 2020.

47. Lombard and Tubiana, "Bringing the Tracker-Guards Back In"; also see White, "Beyond Iraq."

48. Interview with 2020 04_13 WP 1.20b.

49. Interview with 2020_04_13 WP 1.20a

50. Ibid.

51. Ibid.

52. White, "Soldier, Contractor, Trauma."

53. Massé et al., "Green Militarization." Roe et al., *Beyond Enforcement.*

54. Lunstrum, "Green Militarization."

55. Interview with 2020_04_13 WP 1.20b.

56. Massé, Lunstrum, and Holterman, "Linking Green Militarization."

57. As discussed in chapter 5. Rademeyer, *Killing for Profit.* Neumann, "War for Biodiversity." Massé et al., "Green Militarization."

58. Massé and Lunstrum, "Accumulation by Securitization." Lunstrum, "Capitalism." For a discussion of the intersections between ecotourism-security-conservation in Guatemala, see Devine, "Counterinsurgency Ecotourism."

59. https://www.blackmambas.org, accessed 27 May 2020.

60. *The Guardian,* "Africa's New Elite Force: Women Gunning for Poachers and Fighting for a Better Life," accessed 19 December 2018, https://www.theguardian.com/environment/2017/dec/17/poaching-wildlife-africa-conservation-women-barbee-zimbabwe-elephant-rhino.

61. Brockington, *Celebrity and the Environment.*

62. International Committee of the Red Cross, *Professional Standards for Protection Work Carried Out by Humanitarian and Human Rights Actors in Armed Conflict and Other Situations of Violence,* https://www.icrc.org/en/doc/assets/files/other/icrc_002_0999.pdf, accessed 27 May 2020.

63. Interview with 2020_04_13 WP 1.20b.

64. Ibid.

65. Interview with 2020_04_13 WP 1.20a.

66. U.S. Fish and Wildlife Service Division of International Conservation Combating Wildlife Trafficking Program, FY2016 Summary of Projects, more information at https://www.fws.gov/international/pdf/FY16-CWT-project-summaries.pdf, accessed 19 December 2018. *Smithsonian,*

"Giant Rats," accessed 19 December 2018, https://www.smithsonianmag.com/smart-news/how-giant-rats-could-fight-illegal-wildlife-trade-180960925/.

67. Parker, *Detection and Tracking Dogs,* 6.

68. Ibid., 20.

69. WD4C, home page, accessed 20 December 2018, https://wd4c.org.

70. WD4C, "Zambia Anti-poaching," accessed 20 December 2018, https://wd4c.org/zambiaantipoaching.html.

71. WD4C, home page, accessed 20 December 2018, https://wd4c.org.

72. Paramount Solutions, "Skydiving Dogs," accessed 20 December 2018, http://paramountk9solutions.com/skydiving-dogs-poachers-worst-enemies/.

73. PDSA, "Awards," accessed 20 December 2018, https://www.pdsa.org.uk/what-we-do/animal-awards-programme/pdsa-gold-medal.

74. PDSA, "Awards," accessed 20 December 2018, https://www.pdsa.org.uk/what-we-do/animal-awards-programme/pdsa-dickin-medal.

75. Ichickowitz Foundation, "War on Poaching," accessed 20 December 2018, http://www.ichikowitzfoundation.com/2017/12/25/waging-war-on-the-poaching-trade/. Paramount Solutions, "Skydiving Dogs," accessed 20 December 2018, http://paramountk9solutions.com/skydiving-dogs-poachers-worst-enemies/.

76. Defence Web, "Canine Training," accessed 20 December 2018, http://www.defenceweb.co.za/index.php?option=com_content&view=article&id=37164:paramount-unveils-canine-training-school&catid=7:Industry&Itemid=116.

77. Paramount Solutions, home page, accessed 20 December 2018, http://www.paramountgroup.com.

78. Ichickowitz Foundation, "War on Poaching," accessed 20 December 2018, http://www.ichikowitzfoundation.com/2017/12/25/waging-war-on-the-poaching-trade/.

79. Lunstrum "Green Militarization." Massé et al., "Green Militarization."

80. Parker, *Detection and Tracking Dogs,* 6.

81. Ibid., 10.

EPILOGUE

1. https://populationmatters.org, accessed 20 May 2020.

2. IPBES, *Global Assessment Report on Biodiversity and Ecosystem Services,* 2019. Also see Kallis, *Limits.*

3. https://www.half-earthproject.org, accessed 20 May 2020.

4. Büscher et al., "Half Earth or Whole Earth?"; Büscher and Fletcher, "*Convivial Conservation.*"

5. Kothari, "Eco-Swaraj vs. Eco-Catastrophe."

6. http://compassionateconservation.net, accessed 20 May 2020.

BIBLIOGRAPHY

Acharya, Nikhil, and Arthur Mühlen-Schulte. *The Final Round: Combating Armed Actors, Organized Crime and Wildlife Trafficking*, BICC Policy Brief 3\2016. Bonn: BICC, 2016.

Adams, William M. *Against Extinction: The Story of Conservation.* London: Earthscan, 2004.

———. "Geographies of Conservation II: Technology, Surveillance and Conservation by Algorithm." *Progress in Human Geography* 43 no. 2 (2017): 337–50.

Adams, William M., Ros Aveling, Dan Brockington, Barney Dickson, Jo Elliott, Jon Hutton, Dilys Roe, Bhaskar Vira, and William Wolmer. "Biodiversity Conservation and the Eradication of Poverty." *Science* 306, no. 5699 (2004): 1146–49.

Alfino, Sara, and David L. Roberts. "Code Word Usage in the Online Ivory Trade across Four European Union Member States." *Oryx* 54, no. 4 (2020): 494–98.

Amoore, Louise, and Rita Raley. "Securing with Algorithms: Knowledge, Decision, Sovereignty." *Security Dialogue* 48, no. 1 (2017): 3–10.

Annecke, Wendy, and Mmoto Masubelele. "A Review of the Impact of Militarization: The Case of Rhino Poaching in Kruger National Park, South Africa." *Conservation and Society* 14 no. 3 (2016): 195–204.

Anonymous. "Rosewood Democracy in the Political Forests of Madagascar." *Political Geography* 62 (2018): 170–83.

Arsel, Murat, and Bram Büscher. "Nature™ Inc.: Changes and Continuities in Neoliberal Conservation and Market-based Environmental Policy." *Development and Change* 43, no. 1 (2012): 53–78.

Ashaba, Ivan. "Historical Roots of Militarized Conservation: The Case of Uganda." *Review of African Political Economy* 40 (2020), https://doi.org/10.1080/03056244.2020.1828052.

Asiyanbi, Adeniyi. "A Political Ecology of REDD+: Property Rights, Militarized Protectionism, and Carbonised Exclusion in Cross River. *Geoforum* 77 (2016): 146–56.

Ayari, I., and S. Counsell. *The Human Cost of Conservation in Republic of Congo.* London: Rainforest Foundation UK, 2017.

Ayling, Julie. "Reducing Demand for Illicit Wildlife Products: Crafting a 'Whole-of-Society.' " In *Handbook of Transnational Environmental Crime,* edited by Lorraine Elliott and William H. Schaedla. 346–68. Northampton, MA: Edward Elgar, 2016.

———. "What Sustains Wildlife Crime? Rhino Horn Trading and the Resilience of Criminal Networks." *Journal of International Wildlife Law & Policy* 16, no. 1 (2013): 57–80.

Bachmann, Jan, Colleen Bell, and Caroline Holmqvist. *War, Police and Assemblages of Intervention.* London: Routledge, 2015.

Barbora, Sanjay. "Riding the Rhino: Conservation, Conflicts, and Militarization of Kaziranga National Park in Assam." *Antipode* 49, no. 5 (2017): 1145–63.

Barichievy, C., L. Munro, G. Clinning, B. Whittington-Jones, and G. Masterson. "Do Armed Field-Rangers Deter Rhino Poachers? An Empirical Analysis." *Biological Conservation* 209 (2017): 554–60.

Benjaminsen, Tor A., Koffi Alinon, Halvard Buhaug, Jill Tove Buseth, and Nils Petter Gleditsch. "Does Climate Change Drive Land-Use Conflicts in the Sahel?" *Journal of Peace Research* 49, no. 1 (2012): 97–111.

Bersaglio, Brock. "Green Violence: Market-Driven Conservation and the Reforeignization of Space in Laikipia, Kenya." In *Land Rights, Biodiversity Conservation and Justice,* edited by S. Mollett and T. Kepe. 71–88. Abingdon and New York Routledge, 2018.

Bersaglio, Brock, and Frances Cleaver. "Green Grab by Bricolage: The Institutional Workings of Community Conservancies in Kenya." *Conservation and Society* 16, no. 4 (2018): 467–80.

Beseng, Maurice. "Cameroon's Choppy Waters: The Anatomy of Fisheries Crime in the Maritime Fisheries Sector." *Marine Policy* 108 (2019): 1–10.

Bigger, Patrick, and Benjamin D. Neimark. "Weaponizing Nature: The Geopolitical Ecology of the US Navy's Biofuel Program." *Political Geography* 60 (2017): 13–22.

Bigger, Patrick, and Morgan Robertson. "Value Is Simple. Valuation Is Complex." *Capitalism Nature Socialism* 28, no. 1 (2017): 68–77.

Biggs, Duan, Franck Courchamp, Martin Rowan, and Hugh Possingham. "Legal Trade of Africa's Rhino Horns." *Science* 339, no. 6123 (2013): 1038–39.

Blaikie, Piers M., and Harold C. Brookfield. *Land Degradation and Society.* London: Methuen, 1987.

Bocarejo, Diana, and Diana Ojeda. "Violence and Conservation: Beyond Unintended Consequences and Unfortunate Coincidences." *Geoforum* 69 (2016): 176–83.

Bodmer, Richard E., and Eterzit P. Lozano. "Rural Development and Sustainable Wildlife Use in Peru." *Conservation Biology* 15 (2001): 1163–70.

Boekhout van Solinge, Tim. "Deforestation and Crime in the Brazilian Amazon." *Journal of Police Studies* 1 (2016): 87–110.

Booth, Ken. "Security and Emancipation." *Review of International Studies* 17 (1991): 313–26.

Brandon, Katrina, Kent H. Redford, and Steven Sanderson, eds. *Parks in Peril: People, Politics, and Protected Areas.* Washington, DC: Island Press, 1998.

Brashares, Justin S. S., Briana J. Abrahms, Kathryn J. E. Fiorella, Cheryl E. A. Hojnowski, Ryan A. D. Marsh, Tristan A. Nuñez, Katherine Seto, Lauren Withey, Christopher D. Golden, and Douglas J. McCauley. "Wildlife Decline and Social Conflict." *Science* 345, no. 6195 (2014): 376–78.

Broad, Steven, Teresa Mulliken, and Dilys Roe. "The Nature and Extent of Legal and Illegal Trade in Eildlife." In *The Trade in Wildlife: Regulation for Conservation,* edited by Sara Oldfield. 3–22. London: Earthscan, 2003.

Brockington, Dan. *Celebrity and the Environment: Fame, Wealth and Power in Conservation.* London: Zed Books, 2009.

Brockington, Dan, and Rosaleen Duffy. "Capitalism and Conservation: The Production and Reproduction of Biodiversity Conservation." *Antipode* 42, no.3 (2010): 469–84.

Brockington, Dan, Rosaleen Duffy, and Jim Igoe. *Nature Unbound: Conservation, Capitalism and the Future of Protected Areas*. London: Earthscan, 2008.

Broussard, Giovanni. "The Evolving Role of the United Nations Office on Drugs and Crime in Fighting Wildlife and Forest Crimes." In *Handbook of Transnational Environmental Crime*. edited by Lorraine Elliott and William H. Schaedla. 457–68. Northampton, MA: Edward Elgar, 2016.

Buhaug, Halvard. "Climate Not to Blame for African Civil Wars." *Proceedings of the National Academy of Sciences of the USA* 107 (2010): 16477–82.

Burn, Robert W., Fiona M. Underwood, and Julian Blanc. "Global Trends and Factors Associated with the Illegal Killing of Elephants: A Hierarchical Bayesian Analysis of Carcass Encounter Data." *PLoS ONE* 6, no. 9 (2011): E24165.

Büscher, Bram. "Anti-politics as Political Strategy: Neoliberalism and Transfrontier Conservation in Southern Africa." *Development and Change* 41, no. 1 (2010): 29–52.

———. "Derivative Nature: Interrogating the Value of Conservation in 'Boundless Southern Africa.'" *Third World Quarterly* 31, no. 2 (2010): 259–76.

Büscher, Bram, and Robert Fletcher. "Accumulation by Conservation." *New Political Economy* 20, no. 2 (2015): 1–26.

———. *The Conservation Revolution. Radical Ideas for Saving Nature beyond the Anthropocene*. London: Verso, 2020.

———. "Under Pressure: Conceptualizing Political Ecologies of Green Wars." *Conservation and Society* 16, no. 2 (2018): 105–13.

Büscher, Bram, Wolfram Dressler and Robert Fletcher, eds. *Nature Inc.: Environmental Conservation in the Neoliberal Age*. Critical Green Engagements. Tucson: University of Arizona Press, 2014.

Büscher, Bram, Robert Fletcher, Dan Brockington, Chris Sandbrook, William Adams, Lisa Campbell, Catherine Corson, Wolfram Dressler, Rosaleen Duffy, Noella Gray, George Holmes, Alice Kelly, Elisabeth Lunstrum, Maano Ramutsindela, Kartik

Shanker. "Half-Earth or Whole Earth? Radical Ideas for Conservation and Their Implications." *Oryx* 51, no. 3 (2017): 407–10.

Büscher, Bram, and Maano Ramutsindela. "Green Violence: Rhino Poaching and the War to Save Southern Africa's Peace Parks." *African Affairs* 115 (2016): 1–22.

Büscher, Bram, Sian Sullivan, Katja Neves, Jim Igoe, and Dan Brockington. "Towards a Synthesized Critique of Neoliberal Biodiversity Conservation." *Capitalism Nature Socialism* 23, no. 2 (2012): 4–30.

Buzan, Barry, and Ole Wæver. *Regions and Powers: The Structure of International Security.* Cambridge: Cambridge University Press, 2003.

Buzan, Barry, and Jaap De Wilde. *Security: A New Framework for Analysis.* Boulder, Colo.; London: Lynne Rienner, 1998.

Carlson, Khristopher, Joanna Wright, and Hannah Dönges. *In the Line of Fire: Elephant and Rhino Poaching in Africa.* Small Arms Survey Yearbook. Geneva: Small Arms Survey, 2015.

Castree, Noel. "Neoliberalism and the Biophysical Environment 2: Theorising the Neoliberalisation of Nature." *Geography Compass* 4, no. 12 (2010): 1734–46.

Chaber, Anne-Lise, Sophie Allebone-Webb, Yves Lignereux, Andrew A. Cunningham, and J. Marcus Rowcliffe. "The Scale of Illegal Meat Importation from Africa to Europe via Paris." *Conservation Letters* 3, no. 5 (2010): 317–21.

Challender, Daniel, Douglas MacMillan, and Stuart Harrop. "Towards Informed and Multi-faceted Wildlife Trade Interventions." *Global Ecology and Conservation* 3 (2015): 129–48.

Challender, Daniel, and Douglas MacMillan. "Poaching Is More Than an Enforcement Problem." *Conservation Letters* 7 (2014); 484–94.

CITES. *CITES Strategic Vision, 2008–2020.* Geneva: CITES, 2008.

———. *Elephant Conservation, Illegal Killing and Ivory Trade: Report to the Standing Committee of CITES.* 62nd Meeting of the Standing Committee, Document reference SC62 Doc. 46.1. Geneva: CITES, 2012.

———. *Interpretation and Implementation of the Convention: Species Trade and Conservation Rhinoceroses,* Report of the Secretariat CITES document CoP16 Doc. 54.2 (Rev. 1). Geneva: CITES, 2013.

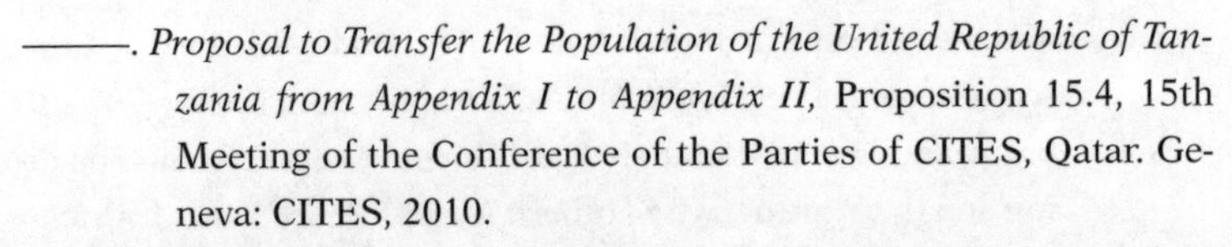

———. *Proposal to Transfer the Population of the United Republic of Tanzania from Appendix I to Appendix II,* Proposition 15.4, 15th Meeting of the Conference of the Parties of CITES, Qatar. Geneva: CITES, 2010.

———. *Proposal to Transfer of the Population of Zambia from Appendix I to Appendix II,* Proposition 15.5, 15th Meeting of the Conference of the Parties of CITES, Qatar. Geneva: CITES, 2010.

———. *Status of Legislative Progress for Implementing CITES,* CoP16 Doc. 28, Annex 2 (Rev. 1). Geneva: CITES, 2013.

Clemente-Munoz, Margarita, A., "The Role of CITES in Ensuring a Sustainable Legal Trade in Wild Flora and Fauna." In *Handbook of Transnational Environmental Crime.* edited by Lorraine Elliott and William H. Schaedla. 433–43. Northampton, MA: Edward Elgar, 2016.

Cochrane, Alasdair, and Steve Cooke. "Humane Intervention: The International Protection of Animal Rights." *Journal of Global Ethics* 12, no. 1 (2016): 106–21.

Commission on Human Security. *Human Security Now: Protecting and Empowering People.* New York: United Nations, 2003.

Cooney, Rosie, and Paul Jepson. "The International Wild Bird Trade: What's Wrong with Blanket Bans?" *Oryx* 40, no. 1 (2006): 18–23.

Cooney, Rosie, Alexander Kasterine, Douglas MacMillan, Simon A. H. Milledge, Katarina Nossal, Dilys Roe, and Michael John't Sas-Rolfes. *The Trade in Wildlife: A Framework to Improve Biodiversity and Livelihood Outcomes.* Geneva: International Trade Centre, UNCTAD, 2015.

Cooney, Rosie, Dilys Roe, Holly Dublin, Jacob Phelps, David Wilkie, and Aiden Keane, et al. "From Poachers to Protectors: Engaging Local Communities in Solutions to Illegal Wildlife Trade." *Conservation Letters* 10, no. 3 (2016): 367–74.

Corson, Catherine A. *Corridors of Power: The Politics of Environmental Aid to Madagascar.* New Haven, CT: Yale University Press, 2016.

Council of the European Union, *Aid for Trade,* Brussels: Council of the European Union, 2007.

Dalby Simon. *Anthropocene Geopolitics: Globalization, Security, Sustainability,* Ottawa: University of Ottawa Press, 2020.

———. *Security and Environmental Change.* Cambridge: Polity, 2009.

Day, Christopher. "Sister Forces: Park Rangers and Regime Security in African States." *Civil Wars* 22, nos. 2–3 (2020): 353–78.

Davis, McKenna, Lucy Smith, Ennid Roberts, et al. *EU Accession to CITES—Main Issues and Positions for the 17th COP* IP/A/ENVI/2016–06. Brussels: European Parliament, 2016.

De Coning, Eve. "Fisheries Crime." In *Handbook of Transnational Environmental Crime,* edited by Lorraine Elliott and William H. Schaedla. 146–67. Northampton, MA: Edward Elgar, 2016.

Demmers, Jolle, and Lauren Gould. "An Assemblage Approach to Liquid Warfare: AFRICOM and the 'Hunt' for Joseph Kony." *Security Dialogue* 49, no. 5 (2018): 364–81.

Dempsey, Jessica, and Morgan Robertson. "Ecosystem Services: Tensions, Impurities, and Points of Engagement within Neoliberalism." *Progress in Human Geography* 36, no. 6 (2012): 758–79.

Deutz, A., G. M. Heal, R. Niu, E. Swanson, T. Townshend, L. Zhu, A. Delmar, A. Meghji, S. A. Sethi, and J. Tobin-de la Puente. *Financing Nature: Closing the Global Biodiversity Financing Gap.* Report by The Paulson Institute, The Nature Conservancy, and the Cornell Atkinson Center for Sustainability, 2020.

Devine, Jennifer. "Counterinsurgency Ecotourism in Guatemala's Maya Biosphere Reserve." *Environment and Planning D: Society and Space* 32 (2014): 984–1001.

Dickinson, Hannah. "Black Gold and Grey Areas: Examining the Impacts of Regulations on the Geopolitical-Ecology of Caviar Trade in the European Union." PhD thesis, University of Sheffield, 2020.

Dickman, Amy, et al. "Wars over Wildlife: Green Militarization and Just War Theory." *Conservation and Society* 18 no. 3 (2020): 293–97.

Di Minin, Enrico, Christoph Fink, Tuomo Hiippala, and Henrikki Tenkanen. "Use of Machine Learning to Investigate Illegal Wildlife Trade on Social Media." *Conservation Biology* 33 (2018): 210–13.

Dressler, Wolfram, Bram Büscher, Michael Schoon, Dan Brockington, Tanya Hayes, Christian A. Kull, James McCarthy, and Krishna Shrestha. "From Hope to Crisis and Back Again? A Critical History of the Global CBNRM Narrative." *Environmental Conservation* 37, no. 1 (2010): 5–15.

Duffield, Mark. *Development, Security and Unending War.* Cambridge: Polity, 2007.

Duffy, Rosaleen. "Global Environmental Governance and North-South Dynamics: The Case of the CITES." *Environment and Planning C: Government and Policy* 31, no. 2 (2013): 222–39.

———. *Killing for Conservation: Wildlife Policy in Zimbabwe.* Oxford: James Currey, 2000.

———. *Nature Crime: How We're Getting Conservation Wrong.* New Haven: Yale University Press, 2010.

———. "Waging a War to Save Biodiversity: The Rise of Militarized Conservation." *International Affairs* 90 no. 4 (2014): 819–34.

———. "War, by Conservation." *Geoforum* 69 (2016): 238–48.

Duffy, Rosaleen, Francis Massé, Emile Smidt, Esther Marijnen, Bram Büscher, Judith Verweijen, Maano Ramutsindela, Trishant Simlai, Laure Joanny, and Elizabeth Lunstrum. "Why We Must Question the Militarization of Conservation." *Biological Conservation* 232 (2019): 66–73.

Duffy, Rosaleen, Freya A. V. St. John, Bram Büscher, and Dan Brockington. "Toward a New Understanding of the Links between Poverty and Illegal Wildlife Hunting." *Conservation Biology* 30, no. 1 (2016): 14–22.

Dunlap, Alexander, and Jostein Jakobsen. *The Violent Technologies of Extraction: Political Ecology, Agrarian Studies and Capitalist World Eater.* London: Palgrave, 2020.

Dutta, Anwesha. "Forest Becomes Frontline: Conservation and Counterinsurgency in a Space of Violent Conflict in Assam, Northeast India." *Political Geography* 77 (2020), https://doi.org/10.1016/j.polgeo.2019.102117.

Dwyer, Michael B., Micah Ingalls, and Ian G. Baird. "The Security Exception: Development and Militarization in Laos's Protected Areas." *Geoforum* 69 (2016): 207–17.

Dzingirai, Vupenyu. "The New Scramble for the African Countryside." *Development and Change* 34, no. 2 (2003): 243–64.

Eckersley, Robyn. "Ecological Intervention: Prospects and Limits." *Ethics and International Affairs* 21, no. 3 (2007): 292–316.

Elliott, Lorraine. "Criminal Networks and Illicit Chains of Custody in Transnational Environmental Crime." In *Handbook of Transnational Environmental Crime,* edited by Lorraine Elliott and William H. Schaedla. 24–44. Northampton, MA: Edward Elgar, 2016.

———. "Governing the International Political Economy of Transnational Environmental Crime." In *Handbook of the International Political Economy of Governance,* edited by Lorraine Elliott and William H. Schaedla. 459–68. Northampton, MA: Edward Elgar, 2014.

———. "The Securitization of Transnational Environmental Crime and the Militarization of Conservation." In *Handbook of Transnational Environmental Crime,* edited by Lorraine Elliott and William H. Schaedla. 68–87. Northampton, MA: Edward Elgar, 2016b.

Elliott, Lorraine, and William H. Schaedla, eds. *Handbook of Transnational Environmental Crime.* Northampton, MA: Edward Elgar, 2016.

Ellis, Stephen. "Of Elephants and Men: Politics and Nature Conservation in South Africa." *Journal of Southern African Studies* 20, no. 1 (1994): 53–69.

Emslie, Richard H., Tom Milliken, and Bibhab Talukdar. "African and Asian Rhinoceroses—Status, Conservation and Trade." In *A Report from the IUCN Species Survival Commission (IUCN/SSC) African and Asian Rhino Specialist Groups and TRAFFIC to the CITES Secretariat Pursuant to Resolution Conf, 9.14,*. compiled by Richard H. Emslie, Tom Milliken, Bibhab Talukdar, Gayle Burgess, Keryn Adcock, David Balfour and Michael H Knight. Gland, Switz.: IUCN, 2012.

Eubanks, Virginia. *Automating Inequality: How High-Tech Tools Profile, Police, and Punish the Poor.* London: St. Martin's Press, 2018.

European Commission, *Action Plan for Strengthening the Fight against Terrorist Financing.* Brussels: European Commission, 2016.

———. *Aid for Trade Report 2015.* Directorate-General for International Cooperation and Development. Brussels: European Commission, 2015.

———. *European Agenda on Security*. Brussels: European Commission, 2015.

———. *The EU Special Incentive Arrangement for Sustainable Development and Good Governance (GSP+) Covering the Period 2014–2015.* Brussels: European Commission, 2016.

———. *EU-Vietnam Free Trade Agreement.* Brussels: European Commission, 2016.

———. *Human Rights and Sustainable Development in the EU- Vietnam Relations with Specific Regard to the EU-Vietnam Free Trade Agreement.* Brussels: European Commission, 2016.

———. *Study on the Interaction between Security and Wildlife Conservation in sub-Saharan Africa.* Brussels: European Commission-DG DEVCO, 2019.

———. *Trade for All: Towards a More Responsible Trade and Investment Policy.* Brussels: European Commission, 2015.

———. *Trade Sustainability Impact Assessment. Free Trade Agreement between the European Union and Japan European Commission/Directorate-General for Trade.* Brussels: European Commission, 2016.

European Union, *EU Action Plan against Wildlife Trafficking,* 2016–2020. Brussels: European Commission, 2016.

EUROJUST, *Strategic Project on Environmental Crime Report.* The Hague: EUROJUST, 2014.

EUROPOL, *Threat Assessment 2013—Environmental Crime in the EU,* Council doc. 15915/13. The Hague: EUROPOL, 2013.

FATF. *Money Laundering and the Illegal Wildlife Trade.* Paris: FATF, 2020.

Felbab-Brown, Vanda. *The Extinction Market; Wildlife Trafficking and How to Counter It.* Washington, DC: Brookings Institute, 2017.

Fischer, A., V. Kereži, B. Arroyo, M. Delibes-Mateos, D. Tadie, A. Lowassa, O. Krange, and K. Skogen. "(De)legitimising Hunting—Discourses over the Morality of Hunting in Europe and Eastern Africa." *Land Use Policy* 32 (2013): 261–70.

Floyd, Rita. "Can Securitization Theory Be Used in Normative Analysis? Towards a Just Securitization Theory." *Security Dialogue* 42, no. 4–5 (2011): 427–39.

Floyd, Rita, and Richard Matthew, eds. *Environmental Security: Approaches and Issues.* London and New York: Routledge, 2013.

Food and Agricultural Organization *International Plan of Action to Prevent, Deter and Eliminate Illegal, Unreported and Unregulated Fishing (IPOA-IUU).* Rome: UN-FAO, 2001.

Fortwangler, Crystal. "Friends with Money: Private Support for a National Park in the US Virgin Islands." *Conservation and Society* 5, no. 4 (2007): 504–33.

Gabrys, Jennifer. *Program Earth: Environmental Sensing Technology and the Making of a Computational Planet.* Minneapolis: University of Minnesota Press, 2016.

Gerstetter, Christiane, Christoph Stefes, Michael Faure, and Niels Philipsen. "Environmental Crime and the EU: Synthesis of the Research Project." European Union Action to Fight Environmental Crime (EFFACE)." Berlin: Ecologic Institute, 2016.

Gleditsch, Nils.P. "Whither the Weather? Climate Change and Conflict." *Journal of Peace Research* 49 (2012): 3–9.

Gore, Meredith L., ed. *Conservation Criminology.* Chichester: John Wiley & Sons, 2017.

———. "Global Risks, Conservation, and Criminology." In *Conservation Criminology,* edited by Meredith Gore, 1–23, 2017.

Gore, Meredith, Patrick Braszak, James Brown, Phillip Cassey, Judith Fisher, Jessica Graham, Ronit Justo-Hanni, Andrea Kirkwood, Elizabeth Lunstrum, Catherine Machalaba, Francis Massé. Maria Manguiat, Delon Omrow, Peter Stoett, Tanya Wyatt and Rob White. "Transnational Environmental Crime Threatens Sustainable Development." *Nature Sustainability* 2 (2019): 784–86.

Grandia, Liza. "Imagining a New Wildlife Politics: Conservation Contrarians and the Corporate Elephants in the Room." *Journal of International Wildlife Law and Policy* 15, no. 1 (2012): 95–114.

Gregory, Derek. "Eyes in the Sky—Bodies on the Ground." *Critical Studies on Security* 6, no. 3 (2018): 347–58.

———. "From a View to a Kill: Drones and Late Modern War." *Theory, Culture & Society* 28, no. 7–8 (2011): 188–215.

Gustafson, Kristian, Touko Sandstrom, and Luke Townsend. "The Bush War to Save the Rhino: Improving Counter-poaching Through Intelligence." *Small Wars & Insurgencies* 29, no. 2 (2018): 269–90.

Haas, T. C., and Samuel M. Ferreira. "Finding Politically Feasible Conservation Policies: The Case of Wildlife Trafficking." *Ecological Applications* 28 no.2 (2018): 473–94.

Haenlein, C., and T. Keatinge. *Follow the Money: Using Financial Investigation to Combat Wildlife Crime.* London: RUSI, 2017.

Hanks, John. *Operation Lock and the War on Rhino Poaching.* Cape Town: Penguin, 2015.

Harrison, Joseph R., David L. Roberts, and Julio Hernandez-Castro. "Assessing the Extent and Nature of Wildlife Trade on the Dark Web." *Conservation Biology* 30, no. 4 (2016): 900–904.

Harrop, Stuart R. "Wild Animal Welfare in International Law: The Present Position and the Scope for Development." *Global Policy* 4, no. 4 (2013): 381–90.

Hartter, Joel, and Abraham Goldman. "Local Responses to a Forest Park in Western Uganda: Alternate Narratives on Fortress Conservation." *Oryx* 45, no. 1 (2011): 60–68.

Hastie, Jo, and Tania McCrea-Steele. *Wanted—Dead or Alive: Exposing Online Wildlife Trade.* London: International Fund for Animal Welfare, 2014.

Hauenstein, Severin, Mrigesh Kshatriya, Julian Blanc, Carsten F. Dormann, and Colin Beale. "African Elephant Poaching Rates Correlate with Local Poverty, National Corruption and Global Ivory Price." *Nature Communications* 10 (2019): 2242.

Haysom, Simone. *Digitally Enhanced Responses: New Horizons for Combatting Illegal Wildlife Trade.* Geneva: Global Transparency Initiative, 2018.

Henk, Dan. "Biodiversity and the Military in Botswana." *Armed Forces & Society* 32 no. 2 (2006): 273–91.

———. "The Botswana Defence Force and the War against Poachers in Southern Africa." *Small Wars & Insurgencies* 16 no. 2 (2005): 170–91.

Hernandez-Castro, Julio, and David Roberts. "Automatic Detection of Potentially Illegal Online Sales of Elephant Ivory via Data Mining." *PeerJ Computer Science* 1, no. 1 (2015), https://doi.org/10.7717/peerj-cs.10.

Hilborn, Ray, Peter Arcese, Markus Borner, Justin Hando, Grant Hopcraft, Martin Loibooki, Simon Mduma, and Anthony R. E. Sinclair. "Effective Enforcement in a Conservation Area." *Science* 314, no. 5803 (2006): 1266.

Hinsley, Amy, Tamsin E. Lee, Joseph R. Harrison, and David L. Roberts. "Estimating the Extent and Structure of Trade in Horticultural Orchids via Social Media." *Conservation Biology* 30, no. 5 (2016): 1038–47.

Hollings, Tracey, Mark Burgman, Mary Van Andel, Marius Gilbert, Timothy Robinson, and Andrew Robinson. "How Do You Find the Green Sheep? A Critical Review of the Use of Remotely Sensed Imagery to Detect and Count Animals." *Methods in Ecology and Evolution* 9, no. 4 (2018): 881–92.

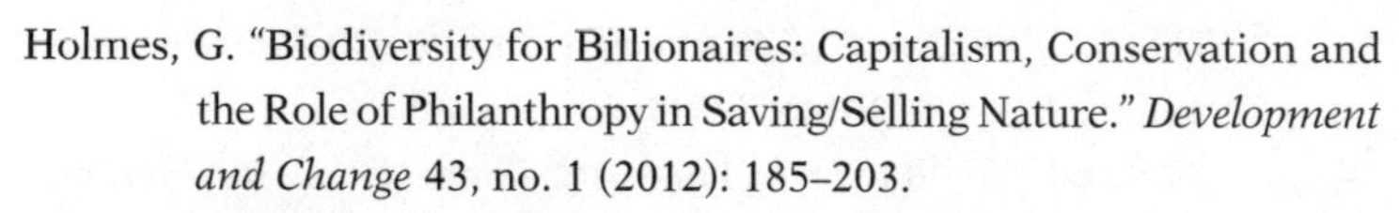

Holmes, G. "Biodiversity for Billionaires: Capitalism, Conservation and the Role of Philanthropy in Saving/Selling Nature." *Development and Change* 43, no. 1 (2012): 185–203.

———. "Conservation's Friends in High Places: Neoliberalism, Networks, and the Transnational Conservation Elite." *Global Environmental Politics* 11, no. 4 (2011): 1–21.

———. "Markets, Nature, Neoliberalism and Conservation through Private Protected Areas in Southern Chile." *Environment and Planning A* 47, no. 4 (2015): 850–66.

Holmes, George, and Connor Cavanagh. "A Review of the Social Impacts of Neoliberal Conservation: Formations, Inequalities, Contestations." *Geoforum* 75 (2016): 199–209.

Homer-Dixon, Thomas. *Environment, Scarcity, and Violence.* Princeton: Princeton University Press, 1991.

———. "Environmental Scarcities and Violent Conflict: Evidence from Cases" *International Security* 19 (1994): 5–40.

———. "On the Threshold: Environmental Changes as Causes of Acute Conflict." *International Security* 16 (1991): 76–116.

Howell, Alison. "The Global Politics of Medicine: Beyond Global Health, against Securitization Theory." *Review of International Studies* 40, no. 5 (2014): 961–87.

Hübschle, Annette. "On the Record: Interview with Major General Johan Jooste (Retired), South African National Parks, Head of Special Projects." *SA Crime Quarterly,* no. 60 (2017): 61–68.

———. "Security Coordination in an Illegal Market: The Transnational Trade in Rhinoceros Horn." *Politikon* 43, no. 2 (2016): 193–214.

———. "The Social Economy of Rhino Poaching: Of Economic Freedom Fighters, Professional Hunters and Marginalized Local People." *Current Sociology* 65, no. 3 (2017): 427–47.

Hübschle, Annette, and C. Shearing. *"Ending Wildilfe Trafficking: Local Communities as Change Agents."* Geneva: Global Initiative against Transnational Organized Crime, 2018.

Humphreys, Jasper, and M. L. R. Smith. "Militarized Responses to the Illegal Wildlife Trade." In *Militarized Responses to Transnational Organized Crime,* edited by Tuesday Reitano, Sasha Jesperson, and Lucia Bird Ruiz-Benitez de Lugo. 25–42. London: Palgrave, 2018.

———. "The 'Rhinofication' of South African Security." *International Affairs* 90, no. 4 (2014): 795–818.

———. "War and Wildlife: The Clausewitz Connection." *International Affairs (Royal Institute of International Affairs 1944–)* 87, no. 1 (2011): 121–42.

Hutton, Jon, William M. Adams, and James C. Murombedzi. "Back to the Barriers? Changing Narratives in Biodiversity Conservation." *Forum for Development Studies* 32, no. 2 (2005): 341–70.

IPBES. *Global Assessment Report on Biodiversity and Ecosystem Services*. Bonn: IPBES, 2019.

IFAW. *Killing with Keystrokes: An Investigation of the Illegal Wildlife Trade on the World Wide Web.* Brussels: IFAW, 2008.

IFAW et al. *Global Wildlife Cybercrime Action Plan.* London: IFAW/INTERPOL/WWF/TRAFFIC/DICE University of Kent/Oxford Martin School, 2018.

Igoe, James. *The Nature of Spectacle: On Images, Money and Conserving Capitalism.* Tucson: University of Arizona Press, 2017.

Igoe, James, and Dan Brockington. "Neoliberal Conservation: A Brief Introduction." *Conservation and Society* 5, no. 4 (2007): 432–49.

Interpol, *Elephant Poaching and Ivory Trafficking in East Africa: Assessment for an Effective Law Enforcement Response.* Lyon: Interpol, Environmental Sub-Directorate, 2014.

IUCN, SULi, IIED, CEED, Austrian Ministry of Environment, and TRAFFIC, *Beyond Enforcement: Communities, Governance, Incentives and Sustainable Use in Combating Wildlife Crime.* Symposium Report, 26–28 February 2015, Glenburn Lodge. Muldersdrift, South Africa, 2015.

Jacoby, Karl. *Crimes against Nature: Squatters, Poachers, Thieves, and the Hidden History of American Conservation.* Berkeley: University of California Press, 2003.

Jahrl, Jutta. "Illegal Caviar Trade in Bulgaria and Romania—Results of a Market Survey on Trade in Caviar from Sturgeons (Acipenseridae)." Vienna: WWF-Austria/TRAFFIC, 2013.

Jeffery, Laura. *Chagos Islanders in Mauritius and the UK: Forced Displacement and Onward Migration.* Manchester: Manchester University Press, 2011.

Jesperson, Sasha. "Assessing Militarized Responses to Transnational Organized Crime." In *Militarized Responses to Transnational Organized Crime.* 1–9. London: Palgrave 2018.

Jiao, Yunbo, Lorraine Elliott, and William H. Schaedla. "Greater China and Transnational Environmental Crime: Understanding Criminal Networks and Enforcement Responses." In *Handbook of Transnational Environmental Crime,* edited by Lorraine Elliott and William H. Schaedla. 255–75. Northampton, MA: Edward Elgar, 2016.

Joanny, Laure. "Law Enforcement Technologies and the Government of Conservation from International Conferences to Indonesian Protected Areas. PhD thesis, University of Sheffield, 2021.

Kashwan, Prakash. *Democracy in the Woods: Environmental Conservation and Social Justice in India, Tanzania and Mexico*. Oxford: Oxford University Press, 2017

Kallis, Giorgis. *Limits: Why Malthus Was Wrong and Why Environmentalists Should Care.* Stanford: Stanford University Press, 2019.

Kassa, Sabba, Jacopo Costa, Robert Lugolobi, and Claudia Baez-Camargo. *A Worm's-eye View of Wildlife Trafficking in Uganda—the Path of Least Resistance.* Basel: Basel Institute on Governance, 2020.

Kasterine, Alexander, et al. *The Trade in South-East Asian Python Skins.* Geneva: International Trade Centre, 2012.

Kelly, Alice B. "The Crumbling Fortress: Territory, Access, and Subjectivity Production in Waza National Park, Northern Cameroon." *Antipode* 47, no. 3 (2015): 730–47.

Kelly, Alice B. Mouadjamou Ahmadou, Mark Moritz, and Paul Scholte. "Not Seeing the Cattle for the Elephants: The Implications of Discursive Linkages between Boko Haram and Wildlife Poaching in Waza National Park, Cameroon." *Conservation and Society* 16, no. 2 (2018): 125–35.

Kelly, Alice B., and A. Clare Gupta. "Protected Areas: Offering Security to Whom, When and Where?" 43, no. 2 (2016): 172–80.

Kelly, Alice B., and Megan Ybarra. "Introduction to Themed Issue: "Green Security in Protected Areas." *Geoforum* 69 (2016): 171–75.

Kleinschmit, Daniela, Stephanie Mansourian, Christoph Wildburger, and Andre Purret, eds. *Illegal Logging and Related Timber Trade: Dimensions, Drivers, Impacts and Responses: A Global Scientific*

Rapid Response Assessment Report. International Union of Forest Research Organizations (IUFRO). Vienna, 2016.

Knapp, Eli J., Nathan Peace, and Lauen Bechtel. "Poachers and Poverty: Assessing Objective and Subjective Measures of Poverty among Illegal Hunters Outside Ruaha National Park, Tanzania." *Conservation and Society* 15, no. 1 (2017): 24–32.

Koh, Lian Pin, and Serge A. Wich. "Dawn of Drone Ecology: Low-Cost Autonomous Aerial Vehicles for Conservation." *Tropical Conservation Science* 5, no. 2 (2012): 121–32.

Kosek, Jake. "Ecologies of Empire: On the New Uses of the Honeybee." *Cultural Anthropology* 25, no. 4 (2010): 650–78.

Kothari, Ashish. "Eco-Swaraj vs. Eco-Catastrophe." *Asia Pacific Perspectives* 15, no. 2 (2018): 49–54.

Kuiper, Timothy, Francis Massé, Nobesuthu A. Ngwenya, Blessing Kavhu, Roseline L. Mandisodza-Chikerema, and Eleanor J. Milner-Gulland. "Ranger Perceptions of, and Engagement with, Monitoring of Elephant Poaching." *People and Nature* (2020), https://doi.org/10.1002/pan3.10154.

Lally, Nick. "Crowdsourced Surveillance and Networked Data." *Security Dialogue* 48, no. 1 (2017): 63–77.

Lavorgna, Anita, C. Rutherford, V. Vaglica, M. J. Smith, and M. Sajeva, M. "CITES, Wild Plants, and Opportunities for Crime." *European Journal on Criminal Policy and Research* 24 (2018): 269–88.

Lavorgna, Anita, and Maurizio Sajeva. "Studying Illegal Online Trades in Plants: Market Characteristics, Organizational and Behavioral Aspects, and Policing Challenges." *European Journal on Criminal Policy and Research* (2020), https://doi.org/10.1007/s10610-020-09447-2.

Lawson, Katherine, and Alex Vines. *Global Impacts of the Illegal Wildlife Trade: The Costs of Crime, Insecurity and Institutional Erosion*. London: Chatham House, 2014.

Leader-Williams, Nigel. "Regulation and Protection: Successes and Failure in Rhinoceros Conservation." In *The Trade in Wildlife: Regulation for Conservation*, edited by Sara Oldfield. 89–99, London: Earthscan, 2003.

Leakey, Richard E. *Wildlife Wars: My Battle to Save Kenya's Elephants*. London: Macmillan, 2001.

Le Billon, Philippe. *Wars of Plunder: Conflicts, Profits and the Politics of Resources.* London and New York: Hurst and Columbia University Press, 2012.

Levi, Michael. "Green with Envy: Environmental Crimes and Black Money." In *Green Crimes and Dirty Money,* edited by Toine Spapens, Rob White, Daan van Uhm, and Wim Huisman. 179–93, Abingdon: Rouledge, 2018.

Lewis, Jerome. "Making the Invisible Visible: Designing Technology for Nonliterate Hunter-Gatherers." In *Subversion, Conversion, Development: Cross-Cultural Knowledge Exchange and the Politics of Design,* edited by James Leach and Lee Wilson. 127–52. Cambridge, MA: MIT Press. 2014.

Li, Ya-Wei, and Timothy Male. *Conservation of Defense: Opportunities to Promote Conservation through Military Readiness.* Washington, DC: Environmental Policy Innovation Center, 2020.

Lindenmayer, David, and Ben Scheele. "Do Not Publish." *Science* 356, no. 6340 (2017): 800–801.

Lipschutz, Ronnie D. *On Security.* New Directions in World Politics. New York: Columbia University Press, 1995.

Litfin, Karen. *Ozone Discourses: Science and Politics in Global Environmental Cooperation.* New York: Columbia University Press, 1994.

Lombard, Louisa. "Threat Economies and Armed Conservation in Northeastern Central African Republic." *Geoforum* 69 (2016): 218–26.

Lombard, Louisa, and Jérôme Tubiana. "Bringing the Tracker-Guards Back In: Arms-Carrying Markets and Quests for Status in Conservation at War." *Political Geography* 77 (2020).

Loperena, Christopher Anthony. "Conservation by Racialized Dispossession: The Making of an Eco-destination on Honduras's North Coast." *Geoforum* 69 (2016): 184–93.

Lubilo, Rodgers, and Paul Hebink. " 'Local Hunting' and Community-Based Natural Resource Management in Namibia: Contestations and Livelihoods." *Geoforum,* 101 (2019): 62–75.

Lunstrum, Elizabeth. "Capitalism, Wealth, and Conservation in the Age of Security: The Vitalization of the State." *Annals of the American Association of Geographers* 108, no. 4 (2018): 1022–37.

———. "Conservation Meets Militarization in Kruger National Park: Historical Encounters and Complex Legacies." *Conservation and Society* 13, no. 4 (2015): 356–69.

———. "Green Militarization: Anti-Poaching Efforts and the Spatial Contours of Kruger National Park." *Annals of the Association of American Geographers* 104 (2014): 1–17.

Lunstrum, Elizabeth, and Nicia Givá. "What Drives Commercial Poaching? From Poverty to Economic Inequality." *Biological Conservation* 245 (2020).

Lunstrum, Elizabeth, and Megan Ybarra. "Deploying Difference: Security Threat Narratives and State Displacement from Protected Areas." *Conservation and Society* 16, no. 2 (2018): 114–24.

Lynch, Michael, and Paul Stretesky. *Exploring Green Criminology: Toward a Green Criminological Revolution.* Burlington, VT: Ashgate, 2014.

Mabele, Mathew Bukhi. "Beyond Forceful Measures: Tanzania's 'War on Poaching' Needs Diversified Strategies More than Militarized Tactics." *Review of African Political Economy* 44, no. 153 (2017): 487–98.

Mackenzie, Catriona, Colin A. Chapman, and Raja Sengupta, "Spatial Patterns of Illegal Resource Extraction in Kibale National Park, Uganda," *Environmental Conservation,* no. 39 (2011): 38–50.

MacKenzie, John. *Empire of Nature.* Manchester: Manchester University Press, 1988.

Maguire, Thomas. J. "Kenya's War on Poaching: Militarized Solutions to a Militarized Problem?" In *Militarized Responses to Transnational Organized Crime: The War on Crime,* edited by Tuesday Reitano, Sasha Jesperson, and Lucia Bird Ruiz-Benitez de Lugo. 61–90. London: Palgrave, 2018.

Maguire, Tom, and Cathy Haenlein. *An Illusion of Complicity: Terrorism and the Illegal Ivory Trade in East Africa.* London: RUSI, 2015.

Margulies, Jared. "The Conservation Ideological State Apparatus." *Conservation and Society* 16, no. 2 (2018): 181–92.

———. "Korean 'Housewives' and 'Hipsters' Are Not Driving a New Illicit Plant Trade: Complicating Consumer Motivations Behind an Emergent Wildlife Trade in *Dudleya farinosa.*" *Frontiers in Ecol-*

ogy and Evolution (2020), https://doi.org/10.3389/fevo.2020.604921.

Margulies, Jared, et al. "Illegal Wildlife Trade and the Persistence of Plant Blindness." *Plants People Planet* 1 (2019): 173–82.

Marijnen, Esther. "The 'Green Militarization' of Development Aid: The European Commission and the Virunga National Park, DR Congo." *Third World Quarterly* 38, no. 7 (2017): 1566–82.

———. "Public Authority and Conservation in Areas of Armed Conflict: Virunga National Park as a 'State within a State' in Eastern Congo." *Development and Change* 49, no. 3 (2018): 790–814.

Marijnen, Esther, Lotje De Vries, and Rosaleen Duffy. "Conservation in Violent Environments: Introduction to a Special Issue on the Political Ecology of Conservation amidst Violent Conflict." *Political Geography* (2020), https://doi.org/10.1016/j.polgeo.2020.102253.

Marijnen, Esther, and Judith Verweijen. "Selling Green Militarization: The Discursive (Re)production of Militarized Conservation in the Virunga National Park, Democratic Republic of the Congo." *Geoforum* 75 (2016): 274–85.

Marris, Emma. "Drones in Science: Fly and Bring Me Data." *Nature* 498 (2013): 156–58.

Martinez-Alier, Joan. *The Environmentalism of the Poor: A Study of Ecological Conflicts and Valuation.* Cheltenham: Edward Elgar, 2002.

Massé, Francis. "Conservation Law Enforcement: Policing Protected Areas." *Annals of the American Association of Geographers* 110, no.3 (2020): 758–73.

———. "Topographies of Security and the Multiple Spatialities of (Conservation) Power: Verticality, Surveillance, and Space-Time Compression in the Bush." *Political Geography* 67 (2018): 56–64.

Massé, Francis, Hannah Dickinson, Jared Margulies, Laure Joanny, Teresa Lappe-Osthege, and Rosaleen Duffy. "Conservation and Crime Convergence? Situating the 2018 London Illegal Wildlife Trade Conference." *Journal of Political Ecology* 27 (2020): 23–42.

Massé, F., A. Gardiner, R. Lubilo, and M. Themba. "Inclusive Anti-poaching? Exploring the Potential and Challenges of Community-based Anti-Poaching." *SA Crime Quarterly* 60 (2017): 19–27.

Massé, Francis, and Elizabeth Lunstrum. "Accumulation by Securitization: Commercial Poaching, Neoliberal Conservation, and the Creation of New Wildlife Frontiers." *Geoforum* 69 (2016): 227–37.

Massé, Francis, Elizabeth Lunstrum, and Devin Holterman. "Linking Green Militarization and Critical Military Studies." *Critical Military Studies* 4, no. 2 (2018): 201–21.

Massé, Francis, and Jared Margulies. "The Geopolitical Ecology of Conservation: The Emergence of Illegal Wildlife Trade as National Security Interest and the Re-shaping of US Foreign Conservation Assistance." *World Development* 132 (2020), https://doi.org/10.1016/j.worlddev.2020.104958.

Matthew, Richard A., Jon Barnett, Bryan McDonald, and Karen L. O'Brien. *Global Environmental Change and Human Security.* Cambridge, MA: MIT Press, 2010.

Mbaria, John, and Mordecai Ogada. *The Big Conservation Lie: The Untold Story of Wildlife Conservation in Kenya.* Auburn, WA: Lens & Pens, 2016.

McAfee, Kathleen. "The Contradictory Logic of Global Ecosystem Services Markets." *Development and Change* 43, no. 1 (2012): 105–31.

———. "Nature in the Market-World: Ecosystem Services and Inequality." *Development* 55, no. 1 (2012): 25–33.

———. "Selling Nature to Save It? Biodiversity and Green Developmentalism." *Environment and Planning D: Society and Space* 17, no. 2 (1999): 133–54.

McAfee, Kathleen, and Elizabeth N. Shapiro. "Payments for Ecosystem Services in Mxico: Nature, Neoliberalism, Social Movements, and the State." *Annals of the Association of American Geographers* 100, no. 3 (2010): 579–99.

McClanahan, Bill, and Tyler Wall. "Do Some Anti-poaching, Kill Some Bad Guys, and Do Some Good': Manhunting, Accumulation, and Pacification in African Conservation." In *The Geography of Environmental Crime: Conservation, Wildlife Crime and Environmental Activism,* edited by Gary R. Potter, Angus Nurse, and Matthew Hall. 121–47. London: Palgrave, 2016.

McDonald, Matthew. "Securitization and the Construction of Security." *European Journal of International Relations* 14, no. 4 (2008): 563–87.

McLellan, Elisabeth, and Crawford Allan. *Wildlife Crime Initiative Annual Update 2015*. Gland, Switz.: WWF and TRAFFIC, 2015.

Mcnamara, J., M. Rowcliffe, G. Cowlishaw, J. S. Alexander, Y. Ntiamoa-Baidu, A. Brenya, and E. J. Milner-Gulland. "Characterizing Wildlife Trade Market Supply-Demand Dynamics." *PLoS ONE* 11, no. 9 (2016): E0162972.

Menon, A., and N. D. Rai. "Putting a Price on Tiger Reserves: Creating Conservation Value or Green Grabbing?" *Economic and Political Weekly* 52 (2017): 23–26.

Milliken, Tom, Richard H. Emslie, and Bibhab Talukdar. *African and Asian Rhinoceroses—Status, Conservation and Trade,* A report from the IUCN Species Survival Commission (IUCN/SSC) African and Asian Rhino Specialist Groups and TRAFFIC to the CITES Secretariat, pursuant to Resolution Conf. 9.14 (Rev. CoP14) and Decision 14.89, CoP15 Doc. 45.1, 2009. Gland, Switz.: IUCN.

Milliken, Tom, Jo Shaw, Richard H. Emslie, Russell D. Taylor, and Chris Turton. *The South Africa–Vietnam Rhino Horn Trade Nexus*. Johannesburg: TRAFFIC, 2012, 134–36.

Millner, Naomi. "As the Drone Flies: Configuring and Vertical Politics of Forests Contestation." *Political Geography* 80 (2020), https://doi.org/10.1016/j.polgeo.2020.102163.

Milner-Gulland, E. J., and Nigel Leader-Williams. "A Model of Incentives for the Illegal Exploitation of Black Rhinos and Elephants: Poaching Pays in Luangwa Valley, Zambia." *Journal of Applied Ecology* 29, no. 2 (1992): 388–401.

Ministry of Defence and Foreign and Commonwealth Office. *UK's International Defence Engagement Strategy*. London: Ministry of Defence, 2017.

Mkutu, Kennedy, and Gerald Wandera. "Policing the Periphery." *Small Arms Survey*. Geneva: Small Arms Survey, 2013.

Mogomotsi, Goemeone E. J., and Patricia Kefilwe Madigelele. "Live by the Gun, Die by the Gun. Botswana's 'Shoot-to-Kill' Policy as an Anti-Poaching Strategy." *South Africa Crime Quarterly* 60 (2017): 51–59.

Mollett, Sharlene, and Kepe, Thembela. *Land Rights, Biodiversity Conservation and Justice: Rethinking Parks and People*. Abingdon and New York: Routledge, 2018.

Moreto, William D. "Introducing Intelligence-Led Conservation: Bridging Crime and Conservation Science." *Crime Science* 4, no. 15 (2015): 1–22.

———. "Occupational Stress among Law Enforcement Rangers: Insights from Uganda." *Oryx* 50, no. 4 (2015): 646–54.

———. *Wildlife Crime: From Theory to Practice.* Philadelphia: Temple University Press, 2018.

Moreto, William D., Rod K. Brunson, and Anthony A. Braga. "Such Misconducts Don't Make a Good Ranger: Examining Law Enforcement Ranger Wrongdoing in Uganda." *British Journal of Criminology* 55, no. 2 (2015): 359–80.

Moreto, William D., Jacinta Gau, Eugene Paoline, Rohit Singh, Michael Belecky, and Barney Long. "Occupational Motivation and Intergenerational Linkages of Rangers in Asia." *Oryx* (2017): 1–10.

Moore, Jason W. *Capitalism in the Web of Life: Ecology and the Accumulation of Capital.* London: Verso, 2015.

Mukamuri, Billy, Jeanette Manjengwa, and Simon Anstey. *Beyond Proprietorship. Murphree's Laws on Community-Based Natural Resource Management in Southern Africa.* Harare: African Books Collective, and Ottawa: Weaver Press, 2009.

Mulero-Pázmány, Margarita, Susanne Jenni-Eiermann, Nicolas Strebel, Thomas Sattler, Juan José Negro, and Zulima Tablado. "Unmanned Aircraft Systems as a New Source of Disturbance for Wildlife: A Systematic Review." *PloS One* 12, no. 6 (2017): E0178448.

Muralidharan, Rahul, and Nitin D. Rai. "Violent Maritime Spaces: Conservation and Security in Gulf of Mannar Marine National Park, India." *Political Geography* 80, no, 6 (2020), https://doi.org/10.1016/j.polgeo.2020.102160.

Murombedzi, James. "Why Wildlife Conservation Has Not Economically Benefitted Communities in Africa." In *African Wildlife and Livelihoods,* edited by D. Hulme and M. Murphree. 208–226. Oxford: James Currey, 2001.

Mushonga, Tafadzwa. "The Militarization of Conservation and Occupational Violence in Sikumi Forest Reserve." *Conservation and Society*, 19, no. 1 (2021): 3–22.

Natusch, Daniel, and Jessica Lyons. "Exploited for Pets: The Harvest and Trade of Amphibians and Reptiles from Indonesian New

Guinea." *Biodiversity and Conservation* 21, no. 11 (2012): 2899–911.

Nellemann, Christian, et al., eds. *The Environmental Crime Crisis—Threats to Sustainable Development from Illegal Exploitation and Trade in Wildlife and Forest Resources.* A UNEP Rapid Response Assessment, United Nations Environment Programme and GRID-Arendal. Nairobi, and Arendal, 2014.

Nellemann, Christian, et al., eds. *The Rise of Environmental Crime—A Growing Threat to Natural Resources Peace, Development and Security.* A UNEP-INTERPOL Rapid Response Assessment, United Nations Environment Programme and RHIPTO Rapid Response–Norwegian Center for Global Analyses, 2016.

Neocleous, Mark. *Critique of Security.* Edinburgh: Edinburgh University Press, 2008.

Neumann, Roderick P. "Moral and Discursive Geographies in the War for Biodiversity in Africa." *Political Geography,* 23 no. 7 (2004): 813–37.

Neumann, Roderick P. "Disciplining Peasants in Tanzania: From State Violence to Self-Surveillance in Wildlife Conservation." In *Violent Environments,* edited by Nancy Lee Peluso and Michael Watts. 305–27. Ithaca: Cornell University Press, 2001.

———. *Imposing Wilderness: Struggles over Livelihood and Nature Preservation in Africa.* California Studies in Critical Human Geography. Berkeley: University of California Press, 1998.

Nijman, Vincent. "An Overview of International Wildlife Trade from Southeast Asia." *Biodiversity and Conservation* 19, no. 4 (2010): 1101–14.

Nijman, Vincent, and Chris Shepherd. "The Role of Asia in the Global Trade in CITES II-Listed Poison Arrow Frogs: Hopping from Kazakhstan to Lebanon to Thailand and Beyond." *Biodiversity and Conservation* 19, no. 7 (2010): 1963–70.

Nurse, Angus, and Tanya Wyatt. *Wildlife Criminology.* Chicago: University of Chicago Press, 2020.

Nuwer, Rachel. *Poached: Inside the Dark World of Wildlife Trafficking.* Boston: Da Capo Press, 2018.

Nyman, Jonna. "What Is the Value of Security? Contextualising the Negative/Positive Debate." *Review of International Studies* 42, no. 5 (2016): 821–39.

O'Brien, Karen, and Jon Barnett. "Global Environmental Change and Human Security." *Annual Review of Environment and Resources* 38, no. 1 (2013): 373–91.

O'Lear, Shannon. *Environmental Geopolitics*. Lanham, MD: Rowman & Littlefield, 2018.

O'Riordan, Tim, and Andrew Jordan. "The Precautionary Principle in Contemporary Environmental Politics." *Environmental Values* 4, no. 3 (1995): 191–212.

Oates, John F. *Myth and Reality in the Rainforest: How Conservation Strategies Are Failing West Africa*. Berkeley: University of California Press, 1999.

OECD. *Illegal Trade in Environmentally Sensitive Goods*. Paris: OECD, 2012.

———. *Illicit Trade: Converging Criminal Networks*. Paris: OECD, 2016.

Parker, Megan. *An Assessment of Detection and Tracking Dogs Programs in Africa*. Working Dogs for Conservation, 2015.

Pattison, James. "Just War Theory and the Privatization of Military Force." *Ethics & International Affairs* 22, no. 2 (2008): 143–62.

———. *The Morality of Private War: The Challenge of Private Military and Security Companies*. 1st ed. Oxford: Oxford University Press, 2014.

Peluso, Nancy L. "Coercing Conservation? The Politics of State Resource Control." *Global Environmental Change* 3 no. 2 (1993): 199–217.

Peluso, Nancy L., and Peter Vandergeest. "Political Ecologies of War and Forests: Counterinsurgencies and the Making of National Natures." *Annals of the Association of American Geographers* 101, no. 3 (2011): 587–608.

Peluso, Nancy L., and Michael.Watts, eds. *Violent Environments*. Ithaca: Cornell University Press, 2001.

Perry, Edward, and Katia Karousakis. *A Comprehensive Overview of Global Biodiversity Finance*. Paris: OECD, 2020.

Phelps, Jacob, Duan Biggs, and Edward L Webb. "Tools and Terms for Understanding Illegal Wildlife Trade." *Frontiers in Ecology and the Environment* 14, no. 9 (2016): 479–89.

Phelps, Jacob L., R. Carrasco, and E. L. Webb. "A Framework for Assessing Supply-Side Wildlife Conservation." *Conservation Biology* 28, no. 1 (2014): 244–57.

Phelps, Jacob L., Edward L. S. Webb, David Bickford, Navjot S. Sodhi, and Vincent Nijman. "Boosting CITES." *Science* 330, no. 6012 (2010): 1752–53.

Phelps Bondaroff, Teale, Wietse van der Werf, and Tuesday Reitano. "The Illegal Fishing and Organized Crime Nexus: Illegal Fishing as Transnational Organized Crime." *Global Initiative against Transnational Organized Crime and the Black Fish* 12, no. 9 (2015).

Pink, Grant. "INTERPOL's NEST: Building Capability and Capacity to Respond to Transnational Environmental Crime." In *Handbook of Transnational Environmental Crime,* edited by Lorraine Elliot and William H. Schaedela. 444-456. Northampton, MA: Edward Elgar, 2016.

Pooley, Simon, John Fa, and Robert Nasi. "No Conservation Silver Lining to Ebola." *Conservation Biology* 29 no. 3 (2015): 965–67.

Pyhälä, Aili, Anna Osuna Orozco, and Simon Counsell. *Protected Areas in the Congo Basin: Failing Both People and Biodiversity?* London: Rainforest Foundation UK, 2016.

Rai, Nitin D., Tor A. Benjaminsen, Siddhartha Krishnan, and C. Madegowda. "Political Ecology of Tiger Conservation in India: Adverse Effects of Banning Customary Practices in a Protected Area." *Singapore Journal of Tropical Geography* 40 no. 1 (2019): 124–39.

Ramutsindela, Maano. "Extractive Philanthropy: Securing Labour and Land Claim Settlements in Private Nature Reserves." *Third World Quarterly* 36, no. 12 (2015): 2259–72.

Ramutsindela, Maano, Marja Spierenburg, and Harry Wels. *Sponsoring Nature: Environmental Philanthropy for Conservation.* London: Earthscan/Routledge, 2011.

Rademeyer, Julian. *Killing for Profit: Exposing the Illegal Rhino Horn Trade.* Cape Town: Penguin Random House South Africa, 2012.

Ratchford, Marina, Beth Allgood, and Paul Todd. "Criminal Nature: The Global Security Implications of the Illegal Wildlife Trade." Washington, DC: International Fund for Animal Welfare (IFAW), 2013.

Reeve, Rosalind, and Steve Jackson. *Policing International Trade in Endangered Species: The CITES Treaty and Compliance.* New York: Earthscan, 2002.

Reitano, Tuesday. "Situating Militarization as Part of an Integrated Response to Organized Crime." In *Militarized Responses to Transnational Organized Crime: The War on Crime,* edited by Tuesday Reitano, Sasha Jesperson, and Lucia Bird Ruiz-Benitez De Lugo. 339–49. London: Palgrave, 2018.

Reitano, Tuesday, Sasha Jesperson and Lucia Bird Ruiz-Benitez De Lugo. *Militarized Responses to Transnational Organized Crime: The War on Crime.* London: Palgrave, 2017.

Rivalan, Philippe, Virginie Delmas, Elena Angulo, Leigh S. Bull, Richard J. Hall, Franck Courchamp, Alison M. Rosser, and Nigel Leader-Williams. "Can Bans Stimulate Wildlife Trade?" *Nature* 447, no. 7144 (2007): 529–30.

Robbins, Paul. *Political Ecology: A Critical Introduction.* 2nd ed. Chichester: John Wiley & Sons, 2011.

Robinson, Janine, Richard Griffiths, Freya A. V. St. John, and David L. Roberts. "Dynamics of the Global Trade in Live Reptiles: Shifting Trends in Production and Consequences for Sustainability." *Biological Conservation* 184 (2015): 42–50.

Rogers, Paul. "Security by 'Remote Control': Can It Work?" *RUSI Journal* 158, no. 3 (2013): 14–20.

Roe, Dilys. "Blanket Bans—Conservation or Imperialism? A Response to Cooney and Jepson." *Oryx* 40, no. 1 (2006): 27–28.

———. "The Origins and Evolution of the Conservation-Poverty Debate: A Review of Key Literature, Events and Policy Processes." *Oryx* 42, no. 4 (2008): 491–503.

———. *Trading Nature: A Report with Case Studies, on the Contribution of Wildlife Trade Management to Sustainable Livelihoods and the Millennium Development Goals.* Cambridge: TRAFFIC-International and WWF-International, 2008.

Roe, Dilys, et al. *Beyond Enforcement: Engaging Communities in Tackling Wildlife Crime.* London: IIED, 2015.

———. *Conservation and Human Rights: The Need for International Standards.* London: IIED, 2010.

Royal Society. *Science: Tackling the Illegal Wildlife Trade.* London: Royal Society, 2018.

Runhovde, Siv. "Merely a Transit Country? Examining the Role of Uganda in the Transnational Illegal Ivory Trade." *Trends in Organized Crime* 21, no. 3 (2018): 215–34.

Sacré, Vinciane. "EU—TWIX: Ten Years of Information Exchange and Cooperation between Wildlife Law Enforcement Officials in Europe." In *Handbook of Transnational Environmental Crime,* edited by Lorraine Elliot and William H. Schaedla. 478–88. Northampton, MA: Edward Elgar, 2016.

Sandbrook, Chris. "The Social Implications of Using Drones for Biodiversity Conservation." *Ambio* 44, no. 4 (2015): 636–47.

Sandbrook, Chris, Rogelio Luque-Lora, and William M. Adams. "Human Bycatch: Conservation Surveillance and the Social Implications of Camera Traps." *Conservation and Society* 16, no. 4 (2018): 493–504.

Sanderson, Steven E., and Kent H. Redford. "Contested Relationships between Biodiversity Conservation and Poverty Alleviation." *Oryx* 37, no. 4 (2003): 389–90.

Selby, Jan, Omar Dahi, Christiane Fröhlich, and Mike Hulme. "Climate Change and the Syrian Civil War Revisited: A Rejoinder." *Political Geography* 60 (2017): 253–55.

Selby, Jan, and Clemens Hoffmann. "Rethinking Climate Change, Conflict and Security." *Geopolitics* 19, no. 4 (2014): 747–56.

Sellar, John M. M. " 'A Blunt Instrument': Addressing Criminal Networks with Military Responses, and the Impact on Law Enforcement and Intelligence." In *Militarized Responses to Transnational Organized Crime: The War on Crime,* edited by Tuesday Reitano, Sasha Jesperson, and Lucia Bird Ruiz-Benitez de Lugo. 91–105. London: Palgrave, 2018.

———. *The UN's Lone Ranger: Combating International Wildlife Crime.* Dunbeath: Whittles Publishing, 2014.

Shaw, Mark. "Soldiers in a Storm: Why and How Do Responses to Illicit Economies Get Militarized?" In *Militarized Responses to Transnational Organized Crime: The War on Crime,* edited by Tuesday Reitano, Sasha Jesperson, and Lucia Bird Ruiz-Benitez de Lugo. 91–105. London: Palgrave, 2018.

Shrestha, Yashaswi, and Renaud Lapeyre. "Modern Wildlife Monitoring Technologies: Conservationists versus Communities? A Case Study: The Terai-Arc Landscape, Nepal." *Conservation and Society* 16, no. 1 (2018): 91–101.

Sina, S., et al. *Wildlife Crime,* Study for the ENVI Committee, IP/A/ENVI/2015-10. Brussels: European Parliament. 2016.

Sollund. Ragnhild A. *The Crimes of Wildlife Trafficking: Issues of Justice, Legality and Morality.* Abingdon: Routledge, 2019.

———. "Wildlife Management, Species Injustice and Ecocide in the Anthropocene." *Critical Criminology* 28 (2020): 351–69.

Sollund, Ragnhild, and Jennifer Maher. *The Illegal Wildlife Trade: A Case Study Report on the Illegal wildlife Trade in the United Kingdom, Norway, Colombia and Brazil.* Oslo and Cardiff: University of Oslo and University of South Wales, 2015.

Somerville, Keith. *Ivory: Power and Poaching in Africa.* Oxford: Oxford University Press, 2017.

Sundström, Aksel, Amanda Linell, Herbert Ntuli, Martin Sjostedt, and Meredith. L. Gore. "Gender Differences in Poaching Attitudes: Insights from Communities in Mozambique, South Africa and Zimbabwe Living Near the Great Limpopo." *Conservation Letters* 13, no.9 (2019), https://doi.org/10.1111/conl.12686.

Soulé, Michael E. "What Is Conservation Biology?" *BioScience* 35, no. 11 (1985): 727–34.

South, Nigel, and Tanya Wyatt. "Comparing Illicit Trades in Wildlife and Drugs: An Exploratory Study." *Deviant Behavior* 32, no. 6 (2011): 538–61.

Spapens, Toine, Rob White, Daan Van Uhm, and Wim Huisman, eds. *Green Crimes and Dirty Money.* Abingdon: Routledge, 2018.

Spierenburg, Marja, and Harry Wels. "Conservative Philanthropists, Royalty and Business Elites in Nature Conservation in Southern Africa." *Antipode* 42, no. 3 (2010): 647–70.

Sullivan, Sian. "Banking Nature? The Spectacular Financialization of Environmental Conservation." *Antipode* 45, no. 1 (2013): 198–217.

Sumaila, Ussif Rashid, Jennifer Jacquet, and Allison Witter. "When Bad Gets Worse: Corruption and Fisheries." In *Corruption, Natural Resources and Development: From Resource Curse to Political Ecology,* edited by Aled Williams and Philippe Le Billon. 93–105, Northampton, MA: Edward Elgar, 2017.

Sumaila, U. Rashid, Vicky W. Y. Lam, Dana D. Miller, Louise Teh, Reg A. Watson, Dirk Zeller, William W. L. Cheung, Isabelle M. Côté, Alex D. Rogers, Callum Roberts, Enric Sala, and Daniel Pauly. "Winners and Losers in a World Where the High Seas Is Closed to Fishing." *Scientific Reports* 5 (2015): 8481.

Suhrke, Astri. "Human Security and the Interests of States." *Security Dialogue* 30, no. 3 (1999): 265–76.

Stiles, Daniel. "The Ivory Trade and Elephant Conservation." *Environmental Conservation* 31, no. 4 (2004): 309–21.

Terborgh, John. *Requiem for Nature.* Washington, DC: Island Press, 1999.

Titeca, Kristof. "Illegal Ivory Trade as Transnational Organized Crime? An Empirical Study into Ivory Traders in Uganda." *British Journal of Criminology* 59, no. 1 (2019): 24–44.

———. "Out of Garamba, into Uganda. Poaching and Trade of Ivory in Garamba National Park and LRA-Affected Areas in Congo." *Analysis and Policy Brief,* no.5. Institute of Development Policy and Management, University of Antwerp, 2013.

———. "Understanding the Illegal Ivory Trade and Traders: Evidence from Uganda." *International Affairs* 94, no. 5 (2018): 1077–99.

Titeca, Kristof, and Theophile Costeur. "An LRA for Everyone: How Different Actors Frame the Lord's Resistance Army." *African Affairs* 114, no. 454 (2015): 92–114.

Titeca, Kristof, and Patrick Edmond. "Outside the Frame: Looking Beyond the Myth of Garamba's LRA Ivory-Terrorism Nexus." *Conservation and Society* 17, no. 3 (2019): 258–69.

Titeca, Kristof, Patrick Edmond, Gauthier Marchais and Esther Marijnen. "Conservation as a Social Contract in a Violent Frontier: The Case of (Anti-) Poaching in Garamba National Park, Eastern DR Congo." *Political Geography* 78 (2020), https://doi.org/10.1016/j.polgeo.2019.102116.

Tittensor, Derek P., Michael Harfoot, Claire McLardy, Gregory L. Britten, Katalin KecseNagy, Bryan Landry, Willow Outhwaite, Becky Price, Pablo Sinovas, Julian Blanc, Neil D. Burgess, Kelly Malsch. "Evaluating the Relationships between the Legal and Illegal International Wildlife Trades." *Conservation Letters,* 2020, https://doi.org/10.1111/conl.12724.

To, Phuc Xuan, Sango Mahanty, and Wolfram Dressler. "Social Networks of Corruption in the Vietnamese and Lao Cross-Border Timber Trade." *Anthropological Forum* 24, no. 2 (2014): 1–21.

Todd, Matthew. *Trade in Malagasy Reptiles and Amphibians in Thailand.* TRAFFIC Southeast Asia, 2011.

TRAFFIC-International, *Recent Significant Wildlife Crime-Related Seizures and Prosecutions, 1997–2019,* Cambridge: TRAFFIC-International, 2019.

———. *What's Driving the Wildlife Trade? A Review of Expert Opinion on Economic and Social Drivers of the Wildlife Trade and Trade Control Efforts in Cambodia, Indonesia, Lao PDR, and Vietnam.* Cambridge, UK/Washington DC: TRAFFIC/World Bank, 2008.

TRAFFIC-International/WWF-International. *Big Wins in the War against Wildlife Crime.* Cambridge, UK/ Gland, Switz.: TRAFFIC-International/WWF-International, 2013.

T'Sas-Rolfes, Michael. "The Rhino Poaching Crisis: A Market Analysis." 2012. http://www.rhinoresourcecenter.com/pdf_files/133/1331370813.pdf.

Twinamatsiko, Medard, Julia Baker, Mariel Harrison, Mahboobeh Shirkhorshidi, Robert Bitariho, Michelle Wieland, Stephen Asuma, E. J. Milner-Gulland, Phil Franks, and Dilys Roe. *Linking Conservation, Equity and Poverty Alleviation.* London: IIED, 2014.

UNCTAD. *Traceability Systems for a Sustainable International Trade in South East Asian Python Skins,* New York: United Nations, 2014.

UNEP. *Green Customs Guide to Multilateral Environmental Agreements.* Nairobi: UNEP, 2008.

———. *UNEP Yearbook.* Nairobi: UNEP, 2014.

UNEP, CITES, IUCN, and TRAFFIC. *Elephants in the Dust: The African Elephant Crisis: A Rapid Response Assessment.* Nairobi: UNEP, 2013.

UNEP-WCMC. *Review of Trophy Hunting in Selected Species.* Cambridge: UNEP-WCMC, 2014.

United Nations. *Report of the Secretary General on the Activities of the United Nations Regional Office for Central Africa and on the Lord's Resistance Army Affected Areas,* Report reference S/2013/297. New York: United Nations, 2013.

UNODC. *Transnational Organized Crime in Eastern Africa.* Vienna: UNODC, 2013.

———. *World Wildlife Crime Report: Trafficking in Protected Species.* Vienna: UNODC, 2016.

Van Asch, Edward. "The International Consortium on Combating Wildlife Crime" In *Handbook of Transnational Environmental Crime,* edited by Lorraine Elliot and William H. Schaedla. 469–77. Northampton, MA: Edward Elgar, 2016.

Van Uhm, Daan P. "Illegal Trade in Wildlife and Harms to the World." In *Environmental Crime in Transnational Context,* edited by A. C. M. Spapens, R. White, and W. Huismans. 43–66. Farnham: Ashgate 2016.

———. *The Illegal Wildlife Trade: Inside the World of Poachers, Smugglers and Traders.* New York: Springer, 2016.

———. "Wildlife and Laundering: Interaction between the Under and Upper World." In *Green Crimes and Dirty Money,* edited by Toine Spapens, Rob White, Daan van Uhm, and Wim Huisman. 197–214, Abingdon: Routledge, 2018.

Venturini, Sofia, and David L. Roberts. "Disguising Elephant Ivory as Other Materials in Online Trade." *Tropical Conservation Science,* 2020, https://doi.org/10.1177/1940082920974604.

Verweijen, Judith. "A Microdynamics Approach to Geographies of Violence: Mapping the Kill Chain in Militarized Conservation Areas." *Political Geography* 79 (2020): 102–53.

Verweijen, Judith, and Esther Marijnen. "The Counterinsurgency/Conservation Nexus: Guerrilla Livelihoods and the Dynamics of Conflict and Violence in the Virunga National Park, Democratic Republic of the Congo." *Journal of Peasant Studies* 45, no. 2 (2018): 300–320.

Vira, Varun, and Thomas Ewing. "Ivory's Curse: The Militarization and Professionalization of Poaching in Africa." Washington DC: Born Free Foundation/C4ADS 2014.

Wainwright, Joel. *Decolonizing Development: Colonial Power and the Maya.* 1st ed. Antipode Book Ser. Hoboken: John Wiley & Sons, 2008.

Wall, Tyler. "Ordinary Emergency: Drones, Police, and Geographies of Legal Terror." *Antipode* 48, no. 4 (2016): 1122–39.

Wanner, Thomas. "The New 'Passive Revolution' of the Green Economy and Growth Discourse: Maintaining the 'Sustainable Development' of Neoliberal Capitalism." *New Political Economy* 20, no. 1 (2014): 1–21.

Wasser, S. K., L. Brown, C. Mailand, S. Mondol, W. Clark, C. Laurie, and B. S. Weir. "Genetic Assignment of Large Seizures of Elephant

Ivory Reveals Africa's Major Poaching Hotspots." *Science* 349, no. 6243 (2015): 84–87.

WCO, *Declaration on the Illegal Wildlife Trade*. Brussels: WCO, 2014.

Wearn, Oliver, Robin Freeman, and David Jacoby. "Responsible AI for Conservation: *Nature Machine Intelligence* 1 (2019): 72–73.

White, Natasha. "The 'White Gold of Jihad': Violence, Legitimization and Contestation in Anti-poaching Strategies." *Journal of Political Ecology* 21, no. 1 (2014): 452–74.

White House, *National Strategy for Combating Wildlife Trafficking*. Executive Order 13648 of July 1, 2013, *Federal Register* 78(129). Washington, DC: US Government, 2014.

Williams, Paul, D. "After Westgate: Opportunities and Challenges in the War against Al Shabaab." *International Affairs* 90, no. 4 (2014): 907–23.

Wilkinson, Rorden, Erin Hannah, and James Scott. "The WTO in Nairobi: The Demise of the Doha Development Agenda and the Future of the Multilateral Trading System." *Global Policy* 7, no. 2 (2016): 247–55.

Williams, Aled, and Philippe Le Billon. *Corruption, Natural Resources and Development: From Resource Curse to Political Ecology*. Cheltenham, UK, and Northampton, MA: Edward Elgar, 2017.

Wilson, Lauren, and Rachel Boratto. "Conservation, Wildlife Crime, and Tough-on-Crime Policies: Lessons from the Criminological Literature." *Biological Conservation* 251 (2020), https://doi.org/10.1016/j.biocon.2020.108810.

Wittemyer, George, Joseph M. Northrup, Julian Blanc, Iain Douglas-Hamilton, Patrick Omondi, and Kenneth P. Burnham. "Illegal Killing for Ivory Drives Global Decline in African Elephants." *Proceedings of the National Academy of Sciences* 111, no. 36 (2014): 13117–21.

Witter, R. "Why Militarized Conservation May Be Counter-Productive: Illegal Wildlife Hunting as Defiance." *Journal of Political Ecology* 28 (1)(2021): 175–92.

Wittig, Tim. "Poaching, Wildlife Trafficking and Organized Crime." *Whitehall Papers* 86, no. 1 (2016): 77–101.

———. *Understanding Terrorist Finance*. Basingstoke: Palgrave, 2011.

Wright, Elisson, Hasita Bhammar, Ana Maria González-Velosa, and Claudia Sobrevila. "Analysis of International Funding to Tackle

Illegal Wildlife Trade." Washington, DC: World Bank Group, 2016.

WTO and CITES. *CITES and the WTO: Enhancing Co-Operation for Sustainable Development.* Geneva: WTO/CITES, 2015.

WWF-International. *Embedding Human Rights in Nature Conservation: From Intent to Action.* Report of the Independent Panel of Experts of the Independent Review of Allegations Raised in the Media Regarding Human Rights Violations in the Context of WWF's Conservation Work. Gland, Switz.: WWF-International, 2020.

WWF-International/TRAFFIC-International. *Big Wins in the War against Wildlife Crime.* Gland, Switz.: WWF-International, 2013.

Wyatt, Tanya. "Exploring the Organization of Russia's Far East's Illegal Wildlife Trade: Two Case Studies of the Illegal Fur and Illegal Falcon Trades." *Global Crime* 10, no. 2 (2009): 144–54.

———. "How Corruption Enables Wildlife Trafficking." In *Corruption, Natural Resources and Development: From Resource Curse to Political Ecology,* edited by Aled Williams and Philippe Le Billon. 154–62. Northampton, MA: Edward Elgar, 2017.

———. "The Illegal Trade of Raptors in the Russian Federation." *Contemporary Justice Review* 14, no. 2 (2011): 103–23.

———. "The Uncharismatic and Unorganized Side to Wildlife Smuggling." In *Handbook of Transnational Environmental Crime,* edited by Lorraine Elliot and William H. Schaedla. 129–45. Northampton, MA: Edward Elgar., 2016.

———. *Wildlife Trafficking: A Deconstruction of The Crime, The Victims and The Offenders,* Basingstoke: Palgrave, 2013.Wyatt, Tanya, and Anh Ngoc Cao. "Corruption and Wildlife Trafficking." *U4 Issue.* Bergen, 2015.

Wyatt, T., and Arielle Kushner. *Global Illicit Wildlife Product Trafficking: Actors, Activities, and Assessment.* National Consortium for the Study of Terrorism and Responses to Terrorism, University of Maryland, 2014.

Wyatt, Tanya, Daan van Uhm, and Angus Nurse. "Differentiating Criminal Networks in the Illegal Wildlife Trade: Organized, Corporate and Disorganized Crime." *Trends in Organized Crime,* 19, no. 1 (2020): 42–66.

Wyler, Liana S., and Pervaze A. Sheikh. *International Illegal Trade in Wildlife: Threats and US Policy.* Report RL34395. Washington, DC: Congressional Research Service, 2013.

Ybarra, Megan. " 'Blind Passes' and the Production of Green Security through Violence on the Guatemalan Border." *Geoforum* 69 (2016): 194–206.

———. *Green Wars: Conservation and Decolonization in the Maya Forest.* Berkeley: University of California Press, 2017.

———. "Taming the Jungle, Saving the Maya Forest: Sedimented Counterinsurgency Practices in Contemporary Guatemalan Conservation." *Journal of Peasant Studies* 39, no. 2 (2012): 479–502.

Yu, Xiao, and Wang Jia. "Moving Targets: Tracking Online Sales of Illegal Wildlife Products in China." *TRAFFIC Briefing*. Cambridge: TRAFFIC-International 2015.

Zain, Sabri. *Behaviour Change We Can Believe in: Towards a Global Demand Reduction Strategy for Tigers*. Cambridge: TRAFFIC-International, 2012.

INDEX